Praise for
Unseen Beings

Unseen Beings is a profound and much-needed book that takes us beyond the important but numbing statistics of what's occurring on our planet. Erik Jampa Andersson points out the underlying causes – our anthropocentrism and lack of awareness of the vital roles that *unseen*, but very much present, beings play in our lives – and the peril of ignoring them. He guides us toward practical ways to shift from our dualistic, selfish, human-centered fixation to include the interconnected web of all beings, seen and unseen. There is no more essential shift than this, and Erik Jampa Andersson has written a brilliant book that gives us a road map to true wholeness and recovery.

LAMA TSULTRIM ALLIONE, AUTHOR OF *WISDOM RISING: JOURNEY INTO THE MANDALA OF THE EMPOWERED FEMININE* AND FOUNDER OF TARA MANDALA

UNSEEN BEINGS

UNSEEN BEINGS

How We Forgot the World Is More Than Human

ERIK JAMPA ANDERSSON

HAY HOUSE

Carlsbad, California • New York City
London • Sydney • New Delhi

Published in the United Kingdom by:
Hay House UK Ltd, The Sixth Floor, Watson House
54 Baker Street, London W1U 7BU
Tel: +44 (0)20 3927 7290; Fax: +44 (0)20 3927 7291
www.hayhouse.co.uk

Published in the United States of America by:
Hay House Inc., PO Box 5100, Carlsbad, CA 92018-5100
Tel: (1) 760 431 7695 or (800) 654 5126
Fax: (1) 760 431 6948 or (800) 650 5115; www.hayhouse.com

Published in Australia by:
Hay House Australia Ltd, 18/36 Ralph St, Alexandria NSW 2015
Tel: (61) 2 9669 4299; Fax: (61) 2 9669 4144; www.hayhouse.com.au

Published in India by:
Hay House Publishers India, Muskaan Complex,
Plot No.3, B-2, Vasant Kunj, New Delhi 110 070
Tel: (91) 11 4176 1620; Fax: (91) 11 4176 1630; www.hayhouse.co.in

Text © Erik Jampa Andersson, 2023

A note on Tibetan language usage: In addition to a number of more familiar languages, Tibetan terms will be encountered throughout this book. On their first occurrence, they are presented with a phonetic rendering, followed by Wylie transliteration (Wyl.), the standard scheme for precisely transliterating Tibetan using Roman characters. On subsequent occurrences, only the phonetic rendering is used. Personal names are only presented in phonetic form.

The moral rights of the author have been asserted.

A catalogue record for this book is available from the British Library.

Tradepaper ISBN: 978-1-78817-809-9
E-book ISBN: 978-1-78817-811-2
Audiobook ISBN: 978-1-78817-810-5

Interior images: i, 1, 75, 123, 149: Unsplash

MIX
Paper from
responsible sources
FSC
www.fsc.org   FSC® C013056

Printed and bound in Great Britain by
TJ Books Limited, Padstow, Cornwall

This book is dedicated to Starr,
the best person I've ever known,
and all those who have been unseen.

CONTENTS

ACKNOWLEDGMENTS

Writing a book is truly no simple task, and *Unseen Beings* never would have come about without the tireless support of my teachers, friends, family, and publishing team.

First and foremost, I must thank my teachers, especially Lama Tsultrim Allione, Dr Nida Chenagtsang, Dr Phuntsog Wangmo, Khenpo Urgyen Wangchuk, and Drubpön Lama Karma, whose instructions and guidance through the years have been utterly transformative. I also want to thank Dr Ronit Yoeli-Tlalim for revolutionizing my understanding of history, and Dr Martin Savransky for inspiring me to think critically about ecology.

To my parents, Edmund and Angela Andersson, I owe a tremendous debt of gratitude for supporting me on my (rather unconventional) path through the years, and for instilling within me the fortitude to ask questions, follow my heart, and live my truth.

My mother also deserves a very special *thank-you* for giving me my first copy of *The Hobbit* 22 years ago. I took a chance on the book, despite its silly-sounding name, only because she had loved it as a child and assured me I would too. The experience proved to be profoundly formative, sparking an enduring passion for myth, language, philosophy, enchantment, and the simple wonders of the natural world. I struggle to imagine what my life would look like today if that gift had never been given, or if my budding passions had not been supported so patiently.

My friends and colleagues have been tremendously supportive through this process. Special thanks go out to Jennifer Storey and Jo Drayton, both of whom spent many hours helping me work through the arguments of this book, as well as Tom Greensmith, Ryan Jacobson, Chandra Easton, Anna Raithel, Eric Rosenbush, Dr Ben Joffe, Christiana Polites, Jo Albright, Matt Mager, and the many others who have inspired me and *been there* for me through the years.

I'm also deeply grateful for the support of my publishing team at Hay House, especially Michelle Pilley and Kezia Bayard-White, who have provided excellent feedback and guidance throughout the writing process, and my wonderfully skillful editor, Lizzie Henry.

Last, but most certainly not least, I want to thank my incredible husband, best friend, and rock, Paolo Valenziano. Your love, counsel, support, and extraordinary patience are truly what allowed this book to come into being. *Grazie mille, amore mio.*

And to Luna and Stella, who have purred at my side from day one to the book's completion, I'm sorry for sometimes needing to use both hands to type, and for neglecting my petting responsibilities. I appreciate your forbearance. Thank you for your companionship and for always reminding me that non-humans can have a *lot* of attitude.

Introduction

THE CALL OF THE UNSEEN

One of the recurring philosophical questions is:
'Does a falling tree in the forest make a sound when
there is no one to hear?' Which says something
about the nature of philosophers, because there is
always someone in a forest. It may only be a badger,
wondering what that cracking noise was.[1]

TERRY PRATCHETT

Why is this a 'recurring' philosophical question? It is because without humans around, we struggle to imagine *any* meaningful experience of 'awareness' in a place like a forest. It's an *environment*, a location, a colorful backdrop for our human dramas, but surely it is *humans* who give it meaning. Even though a falling tree would be sensorily perceived by not only animals, but *even plants and other trees*, many of us still wonder if sound even *exists* without a human around to perceive it.

But the *human-centered story* is killing us. We've become so enraptured by its illusions that we hardly even notice its presence, or the devastating impact it has on the quality of life on Earth.

Not merely for ourselves, but for *all* the countless beings with whom we share the world.

This is a book about transcending anthropocentrism – 'human-centrism,' the belief that *Homo sapiens* are the most important, significant, and 'central' entities in the universe – and creating a better future. But before we can ever hope to move forward with wisdom, we first need the clarity afforded by hindsight. Thus, this book isn't simply a manifesto for change, but a diagnostic exploration of the *roots* of our climate crisis. If we can better understand the 'wrong turns' of the past, we *might* manage to avoid making the same choices in the future. Our future remains *ours* to create – and by *us*, I don't just mean humans.

In reflecting on falling trees, it should be noted that, according to UN estimates, we are losing around 10 million hectares of the world's forests annually – including millions of hectares of tropical primary rainforest.[2] That has a real impact on our world, but even in earnest conversations about climate change, we rarely consider the impact of our actions on other-than-human beings. We instead see so much of what we do as a 'gathering of resources' that has no real connection with 'morality.' We have trained ourselves, over thousands of years, to *unsee* plants and animals as conscious entities. They have no meaning beyond our needs. We may speak of 'saving the planet,' but we usually just mean 'guarding the storehouse,' not 'supporting the welfare of living beings.'

To come to a different way of seeing and behaving, we need more than facts and data. We *already have* the data – we have truly always had it. Long before the first telescope or microscope, we *knew* that we weren't existentially 'alone.' We looked out on the mysterious world around us and simply saw other 'beings.'

But our stories of natural enchantment gradually gave way to *new myths* of human heroes and mighty anthropomorphic gods, the divine allies of a master species destined for lordship over the Earth. This gradual construction of anthropocentric myths slowly drew us into self-obsession. To move toward a brighter future, we must cast these broken myths away. We need *new* stories – or perhaps even *old* ones – to help us truly understand that we are living in a living world, in dynamic *relationship* with all the living beings around us.

Background Influences

I was raised on the works of J.R.R. Tolkien, who first opened my mind to the wonders of *natural enchantment*. Whenever I speak about my 'background,' however briefly, I always begin with this detail, because it aptly demonstrates just how *impactful* stories can be. It was a simple paperback copy of *The Hobbit*, a childhood favorite of my mother's, that first sparked my love of history, languages, myths, plants, cosmology, metalsmithing, and the natural world. I blazed through Tolkien's extended corpus in a matter of a few years and became a relative 'expert' in all things 'Middle-earth.' By 12, I was taking school notes in *tengwar* and dazzling my parents' friends with detailed summaries of tales from the First and Second Ages. I even began offering flowers and singing Elvish poems to Yavanna Kementári, 'Giver of Fruits, Queen of the Earth,' and Varda Elentári, 'the Exalted One, Queen of the Stars,' at a pond near my home in Colorado.[3] While I was sensible enough to realize that Tolkien's tales weren't historically 'factual,' I had a hard time understanding why *The Silmarillion* should be any less important than so-called 'holy books.' I certainly *wanted* Middle-earth to be 'real,' and in a sense I suppose that desire never truly went away – it simply evolved and matured.

In the 20 years since, a lot has changed, and my life has taken many unexpected turns. After struggling with bullying in the public school system, I attended a Christian school for a few years, but I grew to quite dislike theistic religion. I was fascinated by the idea of religion itself, and knew that I wanted to really experience it, but Christianity simply never appealed to me. Nor did creationism, heaven and hell, nor teachings that animals have no souls. I was always particularly sensitive to the last point, given my deep love for our family dog, Starr. I knew I could never follow a faith that thought of animals as soulless objects, so my sights naturally turned eastward.

At 13, I encountered Buddhism, and at 14, I met my root teacher, Lama Tsultrim Allione, and began practicing in the Nyingma tradition of Tibetan Buddhism under her close guidance. I spent many wonderful summers at Tara Mandala, her retreat center in Colorado, receiving extensive training in meditation, philosophy, and ritual arts from some of the greatest teachers of our time. I eventually became the head *umdze* (ritual leader), specializing in the intricate procedures of the *Drupchen* (Great Accomplishment) ceremony and the practice of *Chöd*.

I went on to study religion and Tibetan language at university, but when I was 20 I found myself in a rut. The academic path on which I was set was deeply unfulfilling and I was starting to become rather jaded. I wanted to meaningfully *engage* with the full dynamism of these traditions, not simply poke at them from the outside. The following year, I left university and enrolled in a five-year program in Traditional Tibetan Medicine, *Sowa Rigpa*, the 'Science of Healing.' I received comprehensive training in the root texts and commentaries of Tibetan Medicine, followed by five months living and working in Kathmandu as a medical intern. By 27, I had started my own clinical practice and begun teaching at the request of Dr Nida Chenagtsang. While I was never overly enamored with clinical medicine, I loved teaching and working with ancient texts, and perhaps most of all working with herbs. I became fascinated by plants, not just from the Himalayas, but increasingly from the local countryside and my own backyard. In really 'meeting' these beings, I was beginning to sense that they were far more than useful botanical drugs. I started seeing them as beings, and my relationship with them began to transform from one of 'use' to one of 'relationship.' This shift in thinking was one of the seeds that led to the emergence of this book.

Tibetan Medicine is remarkable for many reasons. It boasts an unbroken lineage of 'integrative' medical research stretching back 800 years, founded upon a sophisticated and syncretic theory of medicine derived from multiple Eurasian traditions. Tibetan doctors today are much as they've always been – rooted in tradition with feet facing forward, constantly seeking to refine and expand their understanding of health, disease, and medical therapies – while often embracing the insights of biomedical science to support and complement their traditional approach to practice.

When Covid-19 hit in 2020, I had just begun seeing patients at my new clinic in London, having moved to the UK after getting married in 2019. With my clinic closed, I had ample time to focus more intently on plants and research, and return to a few projects (like this book) that had been simmering on the back burner. During the early days of the pandemic, I noticed that my colleagues in Tibetan Medicine, Ayurveda, etc., were actively discussing the links between Covid-19 and climate change, while similar conversations were altogether missing from broader public discourse. For traditional doctors, ecological factors weren't a side-consideration, but an integral component of the causes of the crisis itself. But in the public sphere, even dedicated eco-warriors often missed this connection. Instead, many complained that the pandemic was being taken more seriously by global governments than climate change – a valid critique, but only if we pretend that the two issues are entirely unrelated.

It's rare that we acknowledge the ecological dimensions of pandemics, but in Tibetan Medicine, outbreaks of this sort are always contextualized through our relationship with nature. In the *Four Tantras* of Tibetan Medicine, environmentally damaging behaviors like mining, deforestation, and pollution are seen as *provocative* to the more-than-human world, as they negatively impact the *unseen beings* around us. When these beings become ill or are otherwise provoked, a variety of human afflictions can be triggered, including *epidemic infections*.

While working on my MA in History at Goldsmiths College, University of London, I found myself being drawn deeper into the field of Environmental History – the study of human perceptions of, and relations with, the environment over time. Equally, I became fascinated by the intersections between history, science, and mythology, both in the formation of ancient scientific disciplines and in our perceptions of the natural world. Through the study of history, we can better understand not only how ancient systems of knowledge *developed*, but how the visionaries and systematizers behind these developments used *story* and *myth* to explain and authenticate their theories. It has always been through stories that we have come to understand our world, and it is through stories that we can begin to change it.

This book is akin to a tapestry – woven from many different threads of wisdom that I've pulled together over the years. It's part history, part philosophy, part science *and* religion, part eco-critique, part manifesto, and part love letter to the natural world, but at the end of the day, it's a story: a story about how we forgot the world is *more* than 'human' and how our *remembrance* holds the keys to a brighter future for us all.

The Unseen

In setting out to speak of the 'unseen' realm, I am acutely aware that many will expect these pages to be filled with ghosts and supernatural creatures. While some *do* indeed make an appearance, 'unseen beings' refers quite broadly to the many *beings* that we have actively 'unseen.' For much of our history, we have perceived two distinct and opposing worlds: the ensouled world of the 'human' and the inert world of 'nature.' We have come to believe that of the millions of species that inhabit our planet, humans alone possess the magical lucidity of an agentive, rational consciousness – a 'soul,' if you will – while non-humans comprise an unconscious collective that can be broadly classified as 'nature' and 'resources.'

In truth, *unseen beings* can be trees, dandelions, coral reefs, slime molds, even ponds. Even though we can clearly *see* these beings with our naked eyes, we often completely bypass their *being-ness*. We ignore the fact that all of them are engaged in their experience of *life* in an embodied and intelligent way.

Research into non-human and *non-animal* behavior has expanded substantially in recent decades, sparking some grumbling and reshuffling in the scientific world, especially in the field of neuroscience. It's long been assumed that our neurological tissue is *uniquely* capable of producing what we term 'consciousness' or 'awareness.' We know that neurons are integral to *animal* cognition, so their absence in beings like plants and mushrooms has led to the hasty conclusion that they likewise lack the capacity for cognition. But to put it bluntly, this is wrong. We now have an abundance of evidence pointing to the conclusion that even plants, slime molds, and single-celled organisms exhibit behaviors that are indicative of intelligent awareness. It appears irrefutable that *all* life is *fundamentally* aware, and that all beings engage in cognitive processes to accomplish their evolutionary needs.

So, a far more interesting philosophical question than the 'falling tree' problem would be: 'If you chop down a healthy tree for mere entertainment, and it doesn't fall on any humans, does this act have any ethical or moral weight?'

Seeing the Solution

In today's world, our exploitative and dismissive treatment of *unseen beings* is the root cause of countless social and ecological crises. If we want to come to a systematic solution for our existential disease of *climate collapse*, then we need to understand the problem, how we got here, where we're heading, and how to fix it. Accordingly, this book is divided into four sections: *Diagnosis, Causes and Conditions, Prognosis,* and *Treatment.*

In the first section, we'll explore our current circumstances and clarify their departure from a natural state of health. The second section deals with the etiology and pathogenesis of our disorder, investigating the ways that philosophy, religion, and science have all contributed to the development and progression of our disease. The third section deals with our prognosis, especially at the intersection of ecology and illness. We may yet face a grim future, but there are methods of treatment outlined in the fourth section, including adjustments in our philosophies, myths, attitudes, and behaviors, that might guide us toward a meaningful recovery.

This is all, of course, a massive project that warrants *at least* a lifetime of research. I can only hope to scratch the surface. But we don't have lifetimes to wait before decisive changes are made, and none of us knows what lies ahead of us. For me, this book is presented out of a sense of necessity, in a modest effort to try and budge the needle *even a little* in our public discourse. With this approach, we might allow for sprouts of new life to emerge from the cracks of our crumbling world. We can still build a brighter future, but to do so, we must allow ourselves to heed the call of *unseen beings*.

PART I
DIAGNOSIS

1

THE HARD TRUTH

Coming to Terms with
Our Ecological Disease

Our planet is deeply unwell. Most of us already know this to be true. We carry the knowledge as best we can as we go about our lives. We may try not to dwell on it too much, or attempt to channel our anxiety into something 'productive' like activism or volunteer work. We know we need *action*, but what sort of action? Political? Economic? Industrial? Technological? Do we just need to buy electric cars and offset our carbon emissions? Cut down on meat and online shopping? Tighten industrial regulations? Plant some trees?

While all of these approaches could indeed be beneficial, and while they might offer us the satisfaction of feeling we're 'doing something,' we also know that, individually, they're not going to be enough. There is no single pill that can save us from what ails us – our approach to both diagnosis and treatment must be fundamentally more *holistic*.

In themselves, environmental disasters aren't new – nor are the grave effects of human-caused ecological exploitation. Even 2,000 years ago, before an ill-fated sojourn in Pompeii, the Roman historian Pliny the Elder wrote, '... the mind boggles at the thought of the long-term effect of draining the earth's resources and the full impact of greed. How innocent, how happy, indeed how comfortable, life might be if it coveted nothing... except what is immediately available!'[1]

In truth, we have long known we've been heading down a dangerous road, and in many ways this was our intention all along. Our goal has long been to dominate and control the natural world to fashion a *more perfect* human domain for ourselves – to reach a state of 'Peak Humanity.'[2] But a 'peak' is both an apex *and* the beginning of a descent, and it's become unavoidably clear that we may be well past the summit.

Despite the existential danger of our project, we've managed to soothe our fears with a great number of fanciful stories over the years, full of human exceptionalism, divine protection, techno-fixes, and post-apocalyptic salvation. Stories that have convinced us that we're the best, brightest, and most important beings in the universe – and as such, we will always keep progressing. For centuries, the promise of perpetual progress, infinite growth, and a world of limitless opportunities was sold to the so-called 'developed world' under the presumption that the only limitations that could be placed upon us were those that we placed upon ourselves: Where there's a (human) will, there's a way. 'Nature' itself was 'ours' to master.

Over the past half-century, environmental scientists have compiled a staggering amount of research on the 'symptoms' that our planet is presenting, and their prognosis is profoundly clear:

If we carry on with business as usual, we will trigger devastating and irreparable environmental changes leading to widespread ecological and social collapse, catastrophic changes to global climates, and a sixth mass extinction – which is already well underway. If we haven't *already* run out of time to act, the point of no return is quickly approaching.

Usually when we think about mass extinctions, we think principally about the loss of animal life – and indeed in 2020, the World Wildlife Fund (WWF) reported an average 68 per cent decline in animal population sizes over the past 50 years alone.[3] But the loss of life and biodiversity isn't limited to animals. Industrial deforestation claims the lives of around 15 billion trees annually, and this number is rapidly increasing. Every minute, around 300,000 square meters of forest are razed to the ground, largely to clear space for animal agriculture. Rather than logging or papermaking, this is the main driver in the trend, being responsible for around 80 per cent of all industrial deforestation as well as over 50 per cent of overall greenhouse gas emissions.[4]

Due to the evident impotency of our global political establishments, scientists – who have long been taught to remain external to their subject and unbiased in their research – have been forced into the very uncomfortable position of trying to change human behavior *en masse* with data alone. We may think that if we just let climate scientists take the lead, they can safely guide us out of this debacle, but we need a far more intersectional approach. It is not the job of a climate scientist to establish and administer a course of treatment, or to assess the complex social and historical dynamics that caused us to become so bewildered and callous. We do indeed need to *treat the root cause*, but to do so we'll need to begin by acknowledging its complexity.

For all the talk about our planetary disease, we still haven't even agreed on what to *call* it. *Global warming* is, of course, the main topic of most conversations surrounding the crisis. A rapid increase in greenhouse gas emissions triggered by the Industrial Revolution has caused our atmosphere to operate as a heat-trap, driving average global temperatures ever upward and irreversibly destabilizing countless natural ecosystems. From around the 1950s, a 'Great Acceleration' in the global north saw an exponential increase in greenhouse gas emissions, leading to rapid and catastrophic shifts in our global climate. Weather patterns are already beginning to dramatically change, with many regions undergoing devastating droughts, fires, hurricanes, floods, tornadoes, and heat spells unlike anything in recorded history.

But 'global warming' is just one factor in our planet's illness. The Great Acceleration also saw an exponential increase in deforestation, industrial agriculture, and the pollution of the oceans, soil, and waterways – factors that tend to get brushed under the rug when we focus solely on greenhouse gas emissions. But the more inclusive term 'climate change' is equally problematic. Our disease isn't mere *change*. It is a malignant, rapidly spreading cancer. But so far, our approach to mitigating it has been akin to putting a neon bandage on a melanoma. It's time for something different.

Unfortunately, the causes of our crisis span *millennia*, not centuries or decades. They are manifold and complex, but they unavoidably revolve around a foundational root cause – the sundering of human and non-human beings, and our perceptual separation from 'Nature.'

But despite the ancient roots of our crisis, things were very different for most of our history. *Homo sapiens* were, for *at least* 200,000 years, deeply invested in the more-than-human world. We imagined ourselves to be just one node of life in a vast web of agentive beings, and actively pursued relationships with other entities in our shared environment. This only *began* to shift with the widespread adoption of agriculture, and even then its eradication was a slow and selective process.

But over time, *some* societies moved away from this relational model and sought instead to elevate themselves through paradigms of dominance and natural hierarchy, creating cultural environments in which *human* needs, experiences, values, and thoughts were naturally elevated above all other forms of life.[5] Naturally, such hierarchies were based on far more than *species* – we've assigned value based on family, tribal affiliation, ethnicity, sex, gender, appearance, age, physical ability, and a great many other characteristics. Many of these structures still weigh heavy on our modern societies, but *speciesism* and *anthropocentrism* remain uniquely unchecked even in our most 'rational' and 'empirical' fields of inquiry. But our tendency to think only of ourselves in matters of history, science, ethics, industry, art, and philosophy has blinded us to the vital web of life that supports our existence, and indeed to the red flags warning of our own impending doom. The worldview that facilitated our dominance threatens to spell our destruction.

Modern animal agriculture neatly exemplifies, in many ways, the cognitive and emotional dissonance that humans are capable of. We're largely 'animal lovers,' yet we collectively kill more than 100 billion *captive* animals for food, textiles, scientific research, and various other purposes each year – and that's in *addition* to an annual purge of around a trillion wild animals.[6] But the suffering caused by animal agriculture is rarely limited to the lifestock being killed for profit, or the non-humans who were uprooted and evicted to clear the way for more feedlots. The human toll is also catastrophic – from the mass displacement of indigenous peoples and destruction of their ways of life, to the emergence of dangerous novel infections in overcrowded feedlots, industrial animal agriculture has had a devastating effect on nearly all forms of life.

Epidemic infections clearly remain a serious issue in the modern world. Climate scientists and epidemiologists have been warning about the micropathological ramifications of climate disruption for decades, demonstrating time and time again that melting permafrost, changing weather patterns, deforestation, and expanding animal agriculture will invariably expose us to new and dangerous pathogens much like the novel coronavirus. Even with cutting-edge medical advancements like mRNA vaccines and antiviral therapies, we've still lost over 15 million lives to Covid-19.[7] Non-humans have their own ways of maintaining 'natural' boundaries, and it's only with profound hubris that we continue to transgress against them.

Confronting the Anthropocene

Humans are, of course, not magical beings plopped into a terrestrial marketplace full of consumer goods. We are *animals*, very much like the rest, whose unique evolutionary journey has led us to develop a distinct set of skills that has enabled us to do remarkable things. We are indeed *special*, but a chestnut tree is also quite *special*, as is an *Amanita* mushroom, and neither of *them* are actively engaged in the comprehensive destruction of their own environment. At the end of the day, our sense of superiority is nearly always a matter of perspective.

From the point of view of the Earth itself, humanity's only real superlative designation is a profound capacity for destruction. Over the span of our short history, we have gone from being mere creatures of the Earth to formidable 'geological agents,' comparable to the asteroid that ended the reign of the dinosaurs 65 million years ago. In 2000, Paul Crutzen and Eugene Stoermer proposed the term 'Anthropocene' based on evidence that we are in a new geological epoch 'dominated by human activity.'[8]

And yet we don't really use the term to delineate the period through which humans have exerted our power over nature, which reaches back many thousands of years, but rather the present time, in which nature has begun to more loudly fight back. It describes a period in which we can no longer avoid recognizing that nature is *not* a passive object or background, but a vast assembly of interrelated agents, and that we are not above or apart from them, but in fact one *of* them. From locusts to viruses to rivers and oceans, nature is chock-full of 'beings' that can quite systematically undermine any sense of authority that we may have imagined ourselves to possess.[9]

It's also important not to paint 'humanity' with too wide a brush. While I would argue that anthropocentrism lies at the heart of our climate peril, it's important to remember that not all human societies are founded upon anthropocentrism, nor are 'all humans' equally responsible for our climate crisis. As Christophe Bonneuil and Jean-Baptiste Fressoz ask, 'Should the Yanomami Indians, who hunt, fish and garden in the Amazonian forest, working three hours a day with no fossil fuel (and whose gardens have a yield in energy terms nine times higher than the

French farmers of the highly fertile Beauce), feel responsible for the climate change of the Anthropocene?'[10] Clearly not.

Understanding this nuance is important for those of us who actively seek to *de-center* the human in our approach to the world, or to hold humanity accountable for the destruction that 'we' have wrought. At the end of the day, while *human-caused* climate change is a very real and empirical phenomenon, the Anthropocene is still just another kind of story – the incredible tale of how hunter-gatherer apes transformed *en masse* into an environmental force. But thinking that we are something apart from nature and making it an existential 'other' that we can objectify, exploit, or even rescue is itself an integral part of our problem. We must come to realize that we *are* nature – as are all of the animals, plants, microbes, mountains, gases, and weather patterns, and all of the other beings and 'agents' that inhabit and comprise our world. If we can authentically recognize this, then we can undermine the very mentality that has placed us in this crisis to begin with.

The Poison of Anthropocentricity

Identifying a precise starting-point for the Anthropocene is famously contentious. In the early days of the theory, most imagined the epoch to have started around 1800 with the Industrial Revolution. Others saw it as a more recent evolution born out of the Great Acceleration of the past 70 years, while others still place the tipping point in the 'long 16th century,' triggered by European colonial expansion into the Americas and Australasia.[11]

It is understandably complicated to establish a 'primary cause' for such a complex phenomenon. Going back to the disease metaphor, it might be compared to an alcohol-related liver disease or a serious opioid addiction. How can we truly identify the 'starting-point' for such a condition? Would the primary cause be the first drink the patient ever took? Would it be a car accident that led their doctor to put them on prescription painkillers? Or would it be early life traumas? Genetic propensities? Generational dynamics of abuse or disenfranchisement? It turns out that finding a root cause for disease can actually be quite complicated – and ecological disease is no exception.

For many of us, anthropocentricity is so deeply engrained in our collective psyche that we tend to believe it to be an 'objective' truth – the exceptionalism of humanity is simply a fact of life. As such, we sometimes forget that the *idea* of anthropocentrism was also an invention – a dark and ancient 'revolution' contributing to the long history of the Anthropocene. We began laying the foundations for a human epoch many thousands of years ago, on seemingly philosophical or religious grounds. But behind the veneer of 'rational' or 'divine' conclusions, notions of human dominance were always propped up by bad science and logical fallacies – not unlike androcentrism, heterosexism and white supremacy. All such paradigms use overblown taxonomies to elevate one group above all others and justify the perpetuation of a social system that favors only one group's needs.

Many ecofeminists have spoken of the deep historical and philosophical connections between the oppression of women and the domination of non-humans. Val Plumwood suggests that *androcentrism, anthropocentrism, ethnocentrism,* and indeed all *-centrisms* are founded upon the same egocentric impulse to establish a 'master identity.'[12]

To start this process, we have to be aware of this tendency. When these toxic myths are reinforced over many generations, they tend to become endemic to the point of undetectability – at least to those who enjoy the privilege of being a part of the established 'master identity.' Anthropocentrism has become so deeply engrained in our approaches to science, history, theology, art, engineering, storytelling, philosophy, ethics, etc., that we hardly see it at all. We have, quite effectively, learned to *unsee* non-humans as 'beings.'

Our western 'lineage' of anthropocentrism has important ancestral roots in the early Greek writings of Plato and Aristotle, who, in establishing a formal paradigm for the divinely rational and immortal 'human soul,' concluded that all non-human forms of life, even our closest animal kin, were fundamentally nothing more than biological machines – *automated*, but in no way meaningfully alive. Non-human beings were reduced to objects and resources, devoid of any value beyond their human utility.

Of course, *othering* and exploitation were not uniquely Greek inventions, and it would be unfair to blame Plato and his peers for all the systems of oppression that have emerged over the hundreds of thousands of years of human history. But there is something uniquely powerful about a formal *philosophy* of disregard, particularly one that was committed to writing and revered as a model of intellectual sophistication for thousands of years.

In fact, othering has always been a powerful social tool, as the clear delineation of an 'out-group' is one of the easiest ways to strengthen the identity of an 'in-group.' The ethic of *loving one's neighbor*, while a largely universal virtue, is strictly dependent upon *whom* we acknowledge as 'a neighbor.' Humans have, over time, latched onto an array of illusory parameters to establish existential boundaries between *us* and *them*, including species, language, generation, class, religion, ethnicity, skin color, national origin, gender, physical ability, education, and sexual orientation. Whatever boundary we erect, our tendency has long been to *dehumanize* and devalue those outside our in-group to justify their exploitation and alienation.

Over time, these *anthropocentric* and *androcentric* philosophical paradigms were absorbed into trans-cultural movements like 'world religions,' which impacted humans and their non-human neighbors on a truly global scale. Greek anthropocentrism was most notably assimilated in Christianity, which sanctified and disseminated such notions far beyond the Greco-Roman cultural sphere. These new formalized *and* hallowed models of disregard ultimately impacted our approaches to everything from governance to modern science, severely limiting the scope of the questions that we allowed ourselves to ask in the pursuit of 'objective' truth. Until quite recently, it was deemed preposterous to scientifically investigate the inner lives of non-humans, even our closest animal kin. While a brave few dared to challenge such prejudicial paradigms, they were reliably met with harsh resistance, and their world-altering discoveries have unfortunately done very little to significantly impact the lived experience of non-human beings in a human-dominated world.

But it must again be clearly said that we aren't in the Anthropocene because *all humans* are anthropocentric, but rather because these dangerous lineages of anthropocentric thought have infected the societies that have most comprehensively dominated and colonized the modern world. Without a baseline of anthropocentrism, the advancements of the Industrial

Revolution may very well have led to very different outcomes. Technology is but a tool – it can be used for good or for ill, depending on our intentions. But technology will not save us now – only relinquishing our anthropocentric worldview can place us on the road to recovery. But so far, we have done everything in our power to avoid this truth.

Failed Therapies

When it comes to our approaches to treatment thus far, the majority have been at best 'precautionary' gestures, such as taking individual accountability for our waste and consumption, switching to green products, getting a more efficient car, etc. These measures have given us the now ubiquitous concept of an individual 'carbon footprint,' intended to demonstrate that we can avert our grim prognosis as long as *we all do our best* to keep our carbon footprint low. No need for further treatment.

This approach wasn't devised by scientists or policy-makers, but by the public relations gurus at Ogilvy & Mather, a PR firm hired by British Petroleum (BP) to find a way to divert negative attention away from the oil and gas industry in climate conversations.[13] Their 'carbon footprint calculator,' which was launched in 2004, enabled us to see all of the ways in which our own individual behaviors contribute to the warming of the planet. By placing the onus directly onto the consumer, this rather ingenious pivot convinced a generation of environmentalists to blame the everyday person, and not corporate greed, for the destruction of the planet.

The accumulation of carbon in the atmosphere is, however, certainly a useful marker for 'staging' our disease. For most of the past few hundred thousand years, atmospheric CO_2 levels fluctuated between around 200–300 parts per million (ppm), with pre-industrial levels floating around 280 ppm – a level which some climate scientists claim is the 'sweet spot' for at least *human* life, since it results in temperatures that are physiologically quite comfortable for modern humans.[14] But in the Great Acceleration, CO_2 emissions rose from the low 300s to around 415 ppm, the current level,[15] in just 72 years. In 2016, a global body of climate scientists agreed that keeping CO_2 levels below 430 ppm would enable us to meet the target of the Paris Agreement and prevent average global temperatures increasing by more than 1.5°C.[16] As it stands, we are well on track to blow past 430 ppm of atmospheric CO_2 in the coming decade.

Focusing on the data has long been the impulse among climate scientists and environmentalists, but that simply helps us to identify the stage of our disease, not the appropriate treatment. That requires far more complicated and collaborative analysis, interpretation, and decision-making across a wide range of fields. The tides have, however, begun to turn somewhat in modern environmental movements, with far more dialogue surrounding the pivotal role that *capitalism* plays in our processes of environmental exploitation.[17] This is clearly identifiable as one of the most dangerous *co-morbidities* that have contributed to our systemic disease, with the profit-driven commodification of 'natural resources' and the systematic exploitation of human and non-human beings producing one of the most pervasive and destructive ideological paradigms ever imagined in human history. But despite the delusion and devastation that it has wrought, we're often told (even by its critics) that capitalism is incurable – perhaps even endemic – and that there is simply no alternative.

Furthermore, in the past 30 years there has been a distinctly reactionary conservative turn in modern politics, born out of fears that 'climate propaganda' will be used to undermine the fabric of capitalistic Euro-American society. Any hint of industrial regulation is routinely characterized as an attack on 'personal freedoms,' a concern that is repeatedly stoked by the private sector, which routinely exploits anthropocentric myths to reiterate human separation from the limitations of nature: 'If the Earth was created by God for *man*, then it would be ludicrous to think that he would allow our flourishing to result in devastation.' Whether we want to admit it or not, this simple myth remains a perennial justification for climate inaction for many people, supported in no small part by corporate propaganda.

This fear of social and economic disruption is certainly not unfounded. Halting the progression of global warming and mass extinction *would*, in fact, necessitate a revolution in our approaches to industry, consumption, and the appropriation and distribution of resources.[18] But few politicians are willing to acknowledge this. The Thatcher-era mantra 'there is no alternative' has become so insidiously rooted that even capitalism's critics struggle to imagine a world beyond it. But we remain a flexible and creative species. When it comes to creating new realities for ourselves, we are indeed magicians – neither intrinsically benevolent nor malevolent, but fundamentally a bit bewildered by our craft.

With a rising culture of misinformation, conspiracy theories, and abject rejection of science, it's very unlikely that a *data-driven* approach to mitigating climate change is going to get us very far. While it's imperative that we continue to hold corporations accountable and support ambitious climate legislation, we *need* to embrace a more holistic approach. Not only a new approach to industry, economics, or politics, but a new approach to making sense of our crisis in the first place.

When attempting any kind of planetary diagnosis and treatment, we must first be willing to face the music. A dissolution of our existing ways of 'seeing' our planet, as well as our place within it, may sound scary and destabilizing, but it is in fact an incredible opportunity for creativity, experimentation, and revolution. In a rather poetic twist of fate, our Anthropocenic disease has presented us with a task that *could* confer upon us a worthy distinction as a truly noble and wise species. But it is only through relinquishing our lust for dominance, abandoning our delusions of exceptionalism, and recovering our sense of embeddedness in the natural world that we can ever hope to live up to our name, *Homo sapiens*, the 'wise ape.'

Patient Compliance: Science and Stories

Patient compliance is another important consideration in the treatment of any disease, since even the most efficacious remedies are useless if a patient won't take them. There is an art and science to encouraging people to do what is best for them, and unfortunately this craft can be easily manipulated for destructive and divisive purposes. But the medium of inspiration *and* manipulation is the same: It is in *stories*, rather than data, that our hearts tend to place the most stock. Even the most compelling science, without an equally compelling *story*, may fade into obscurity.

What does this mean for our current existential crisis? Well, for one, we need to focus *much* more on not only interpreting data but also *translating* it into the language of the human psyche. Those in positions of power already understand this. We're sold smartphones and designer shoes not as *products*, but as *experiences* and *identities* crafted in the mythic world of an amorphous 'brand.' We reaffirm the existence of these soulless myths through our financial engagement, but over time we've completely lost sight of who is driving the narrative.

Like any strong treatment, myths can function as a medicine or poison – we can continue poisoning ourselves with toxic myths concocted under false pretences, or we can heal ourselves with 'truthful' myths that remind us of our fundamental embeddedness in, and relationship with, the living world.

Myths don't require us to diminish the veracity of data or the importance of empirical sciences. We *need* science – but we also need more integral collaboration between the sciences and humanities, especially ecology and the arts. We need new stories to make the data comprehensible, which then allows the data to inspire even *newer* stories, and the process continues *ad nauseam*.

We will never abandon stories, but we *can* make a conscious decision to nurture the ones that support the highest aims of life on Earth, and to dismantle those that oppress and delude us.

This treatment needn't be unpleasant. Relinquishing our lust for dominance won't bring an end to human flourishing. In fact, we have every reason to believe that it will massively *improve* our subjective quality of life. Recovery from our addiction to anthropocentrism *will* require some individual and collective soul-searching, but the results can be nothing short of *enchanting*.

Re-Enchantment: A Therapeutic Process for Our Time

What is *enchantment*? The verb 'to enchant' comes from the Latin *incantare*, literally '[to place] into a spell/song,' i.e. 'to incant,' with *cantare* meaning 'to sing.' But 'enchantment' has a wide range of connotations, both in English and other European tongues. It can indicate a sense of attraction, enjoyment, and inspiration, but also bewilderment, and both positive and negative experiences of 'bewitchment.'

At its core, it is a fundamentally *relational* experience. We are placed under *someone*'s spell, be they human, beast, or otherwise. It's quite apropos that, in proper French, one says *enchanté* when making someone's acquaintance. This basic meeting of minds is the seed of enchantment itself, and when we allow ourselves to embrace the fullness of the living world, the experience of enchantment naturally compels us toward *empathic connection* with a wide array of other beings. It should come as no surprise that communications between human and non-human beings lie at the heart of so many of our global myths and fairy tales. To be able to commune with non-humans in a meaningful way or, as Tolkien once said, to understand 'the proper languages of birds and beasts and trees'[19] is a deeply primal desire. We seem to be hard-wired for engagement with non-humans – it's simply our contrived philosophical myopia that prevents us from making use of it. We are nature communing with itself, and in that simple recognition we can find a wellspring of deep and profound inspiration.

However, 'enchantment' is somewhat of a dirty word in our culture of disenchanted 'reason' and is usually aligned with falsehood and illusion. In *The Myth of Disenchantment*, Jason

Josephson-Storm argues, however, that Euro-American societies have never *truly* abandoned their predilection for enchantment, even the 'great men and women of science.'[20] Groundbreaking researchers like Galileo, Newton, and Marie Curie all earnestly engaged with 'magical' practices and 'supernatural' beliefs alongside their clear dedication to the scientific process, and while we think of them as moving us ineluctably *away* from some 'primitive state' of enchantment, it's unlikely that they would have characterized their work in that way.

Czech philosopher Ernest Gellner has even argued that though disenchantment may have indeed been a necessary condition for modernity as we know it, human beings simply *can't stand* living in a disenchanted world.[21]

Thus, our method of treatment, should we choose to accept it, is fundamentally a process of *deconstruction* and *re-enchantment* – a return to the wondrous ground. It doesn't require us to abandon truth or become absorbed in the 'supernatural.' Quite the opposite, in fact: *Natural enchantment* pushes us to embrace a wider, more dynamic view of *nature* itself and acknowledge the *plurality of experiences* within our universe. With natural enchantment, we are under nobody's spell, we are simply opening our senses to the full symphonic grandeur of the living world. We have every reason to be enraptured by the marvels of the universe, so why should we settle for anything less?

If we allow ourselves to be truly open to the lives and experiences of others, including those currently unseen, this basic sense of enchantment can be a powerful antidote for prejudice, cynicism, and egocentricity. The simple magic of recognizing the personhood of others can truly change the world.

THE NATURAL STATE

Remembering Our
More-than-Human Kin

N*ature* is not a *place*. We often speak of it as if it were a kind of stage or movie set – indeed 'the Great Globe itself'[1] – on which humans are the cast of actors, animals are the props, and the plant-filled landscape is the backdrop. But while the Earth is certainly our abode, it's not just a dwelling with some pretty background foliage – it's a tightly knit *community* of interconnected beings, some seen, many *unseen*, all engaged in their own affairs and with their own experience of reality. We are not the masters of their world, nor they ours. They are our cousins, our neighbors, and sometimes also our enemies, but within every living node of the vast web of life, there's a whole dimension of subjective experience.

If we wish to cultivate a healthier relationship with our environment, we first need to come to understand it in its *natural* state, so we can tread lightly and engage with it in a compassionate way. We shouldn't think that we are the only creatures who *mustn't* engage with other beings, lest we 'interfere' with the natural world.[2] All this does is reiterate our separateness from nature. We need to amplify and advocate for the unheard voices of the world, not artificially recuse ourselves from 'their' domain.

In Tibetan Medicine, we refer to the state of physiological balance as *tamél né-mé*, 'the natural state without disease.' When we're in such a state, our body, mind, and energy function in cooperative equilibrium, allowing us to respond to shifting circumstances with strength and resilience. It's only when we fall *out* of this state, due to an assortment of factors, that we are propelled into a state of *imbalance*, which creates the conditions for disease to arise.

Through understanding the *natural state* of our more-than-human environment, we can both begin to make sense of what it really means to be *alive* and gain some important insight into the basic nature of our existential disease.

Going Beyond Speciesism

The tapestry of biological life on our planet is astonishingly vast. Nearly every nook and cranny of our green Earth is teeming with living beings, and within each of these beings we can find many basic parallels to our own experience. 'Life' can be broadly defined as 'a *unified* and *organized* system or process that *responds* and *adapts* to external stimuli, usually engaging in processes of *growth*, *consumption*, *metabolization*, *excretion*, and *reproduction*.'[3] But even biologists don't always agree on these basic parameters, with some 'beings' (like viruses) challenging our paradigms of 'living' and 'non-living' organisms altogether. While astrobiologists are actively looking for signs of extra-terrestrial life that might be quite unlike our carbon-based family, so far every observable form of life, strictly speaking, is part of our very own clan. We are *all* related – millipedes, slime molds, bacteria, and humans – and between us we can establish a basic biological 'mutuality of being.'⁴

The non-human capacity for subjective awareness has been a hotbed of debate for many millennia. While most of us now accept the relative sentience of at least *some* non-human animals, dichotomies between animal 'instinct' and human 'free will' have left many of us with the sense that non-human animals are instinct-driven automated machines, while humans – alone – are agentive or 'ensouled' creatures. Of course, those of us who have lived with pets have *experienced* non-human sentience first-hand, but our tendency toward *speciesism* leads many of us to think that our dogs and cats are somehow *unique* in the animal world. We like to tell ourselves that some animals are just *more* alive and conscious than others, thus justifying our preferential treatment of companion pets over livestock and wild animals.

Of course, there have always been those who disagree with such views, and in many traditions animals are fully embraced as sentient and conscious beings. But plants are usually quite a different story. Plants are among the most emblematic examples of *unseen beings* in our human-centered world. They are not *unseen* because they're invisible or supernatural, but because we can stare right at them without ever noticing them staring back at us. For thousands of years, even some of the most inclusive and life-affirming philosophical traditions have turned a blind eye to the experience of plants. Naturally, this hasn't been a universal phenomenon, and there remain many indigenous societies that relate to plants and other non-animal beings as living agents;

however their voices have been routinely silenced through centuries of colonialism and Eurocentric bias.

Until the 21st century, it was generally thought that there was no scientific basis for plant sentience. But the tides are beginning to change. Science has shown us that *intelligence*, once imagined as a unique function of the animal (or even human) brain, is in fact a defining characteristic of even the most 'primitive' forms of life. Even single-celled organisms like *Physarum polycephalum* demonstrate the capacity to learn, remember, and solve simple problems without any trace of neurological tissue, or even multiple cells.[5]

So, intelligent life is everywhere, but how can we relate to it? How can we comprehend the decision-making process of a slime mold or a tree if we struggle to understand the inner lives of other animals, or indeed even ourselves? Shouldn't we be avoiding *anthropomorphic* attestations of 'intelligence' and 'awareness' anyway? Philosopher Michael Marder has argued that projecting such 'human' characteristics onto beings like plants is itself a manifestation of our 'narcissistic self-recognition of the human in the environment.'[6] Is it anthropocentric to think that other beings have anything at all in common with us?

While we necessarily have a difficult time making sense of *any* realities outside of our own personal frame of experience, admonitions against anthropomorphism are usually levied to divert attention away from a recognition of the basic *continuities* between the life of humans, animals, and plants.[7]

Of course there is only *one* mode of experience that we can ever truly know from the inside – that of the human. But as a species on planet Earth that has evolved through the same basic processes as all other forms of life, we can hardly be characterized as *outside* observers.

Our Animal Kin

In fact, humans are animals, pure and simple. We speak of our 'animal nature' as if it encompasses only our flaws and shortcomings, but in fact *the entirety of our being* is our 'animal nature.' We are physiologically and cognitively complex *because* animals are physiologically and cognitively complex, and the complexity of animals is itself a reflection of the vital complexity of *all life*. We are only just now beginning to *scientifically* come to terms with the depth of non-human intelligence, and we've quickly discovered that we don't need to go very far to find profoundly clever animals.

Crows, for instance, are likely every bit as intelligent as chimpanzees, with similar capacities for tool-building and problem-solving, and clear comprehension of the minds of others.[8] But even pigeons, frequently regarded as 'rats with wings' in metropolises around the world, are rather unsung geniuses all their own. We've employed them as messengers for at least 3,000 years, and in the 1990s some researchers even managed to train a group of pigeons to successfully distinguish between paintings by Picasso and Monet, even if they were composed in a similar style and the birds had never seen them before.[9]

Of course, discussions about the animal kingdom often inevitably revolve around the 'law of the jungle' and 'survival of the fittest,' as if the world of beasts is merely one of grit and competition. But it's actually *cooperation* that often affords the greatest evolutionary advantages in the various kingdoms of life. Cooperation and mutual care are *fundamental* to existence as we know it, and if we want to better understand the natural world and our place within it, it's helpful to begin with the power of community.

Take the common honeybee. There's been a fortunate boom in insect and bee-appreciation in recent years, sparked by our realization that they're an indispensable part of the complex network of organisms that keep us alive. But bees aren't just some bumbling bugs who pollinate our crops, they are intelligent and social beings. Their sociability is an evolutionary *imperative*, since bee colonies often reach populations of up to 50,000 members. Humans didn't start experimenting with communities of this size until around 5,000 years ago, and to do so we needed to make some significant technological and social strides.

Bee societies are both complex and collaborative. While the queen is certainly treated like a royal, with an entourage of her children tending to her every need, bees make big decisions *democratically*, not by royal decree. When a colony gets too large, a group of some thousands of bees will splinter off and create a new colony. But deciding where to establish a new home is no simple task for these intelligent and opinionated insects. To come to an educated group decision, they set up a temporary camp and send scouts to explore the surrounding areas, relating accounts of their journeys upon their return through the language of interpretive dance. Other bees then set out on a property viewing with the scout whose dance they found most compelling, after which they return and present their *own* dances for those who remain undecided. After many rounds of 'voting,' the location that garners majority support becomes the site of the new colony.[10]

Collaborative decision-making is in fact a pervasive feature of the natural world, expressed through the roots of plants, colonies of insects, packs of wolves, and even the neurological circuitry of the human brain. Multicellular organisms are themselves a testament to the *power of collaboration*. Our ability to make friends, influence people, and make collective decisions is part of what makes us *alive*, not just *human*.

Friends for Life

While collaboration within a homogenous community like a bee colony is perhaps unsurprising, the natural world is also replete with examples of inter-species cooperation. Many such relationships even bridge multiple kingdoms of life, with animals, plants, and fungi working together for the common good.

In our own evolution as a species, non-humans have often played crucial roles. Plants and animals weren't always just our food and possessions – they were our mentors, companions, even our ancestors.

There's one non-human, in particular, whose profound impact on our human story warrants far more recognition. These beings waltzed into our lives around 20–40,000 years ago,[11] though our ancestors had watched them from afar for many generations. They were descended from the beasts of legend – formidable hunters who commanded vast swathes of land with ferocious might. In many of our myths and legends, they were immortalized as guardians of the underworld and crucial intermediaries between the human and non-human domains. Before we had ever tamed a horse, milked a cow, or sown a field of grain, we had befriended a *dog*.[12]

Dogs have been with us *at least* since the peak of our most recent Ice Age (20–27,000 years ago), having evolved from Pleistocene wolves – the common ancestor of both the modern grey wolf (*Canis lupus*) and the dog (*Canis lupus familiaris*). It's believed that humans and wolves were gradually drawn together during the perilously harsh conditions of the Last Glacial Maximum. As our paths began to cross more and more frequently in our pursuit of mutual prey, what likely started as a timid sharing of spoils led to an unusual sense of kinship between the two predators. Wolves were drawn into the warmth of human encampments, and ultimately made themselves quite at home at the foot of our beds. They offered us vital protection, companionship, and a natural 'security alarm' in a wild and dangerous world, while we provided them with warmth, food, and evidently also emotional satisfaction.[13]

Studies of canine intelligence have repeatedly attested to dogs' advanced capacity for memory, social cognition, inferential learning, and even comprehension (and possible use) of human language.[14] But beyond their clear intelligence, what deserves significantly more attention is the very real impact dogs have had on our own evolutionary trajectory. Unlike predators who prefer to prey on weaker animals, wolves thrived as *persistence* hunters, successfully felling giant mammals by stalking them to exhaustion in well-organized packs. As *territorial* animals, they also went to the great trouble of staking out their own tribal domains, maintaining a distinctly *pastoral* lifestyle in complex social groups. Such practices were wholly foreign to early humans and other simians, but by the time our ancestors finally found their footing in the Eurasian wilderness, they had become rather formidable and territorial pack hunters themselves. Researchers suggest that these novel human behaviors were at least partially influenced by our burgeoning relationship with canines, who introduced us to their world, taught us their hunting tricks, and afforded us peace of mind by protecting our settlements against less amiable foes. The domestication of dogs was one of the key forces that led to the development of fully modern humans, impacting our relationships with one another and the world at large for many millennia to come.[15]

The earliest archeological evidence of the domestic dog is the *Bonn-Oberkassel dog*, who was ceremoniously laid to rest beside two human corpses around 14,000 years ago.[16] All the remains in the grave, including the dog's, were dusted with distinctive red ocher, a pigment used widely in ceremonial contexts across the world. This special interspecies burial, like many others found on every inhabited continent, points to a significant *familial* relationship between humans and canines globally. But beyond the burial itself, the fact that dogs were buried *ceremoniously* is of particular interest. Ritual burial, especially with material offerings, is a key marker for notions of an *afterlife*. Dogs, like humans, were often buried with grave goods, as in the case of a Siberian dog interred 9,000 years ago with a large spoon and other goods.[17] It's implied that the dog would have a *need* for such things after death, and also that there would be *other* beings around who could make practical use of something like a spoon, possibly to deliver his evening meal. It's very possible that, to prehistoric humans, the human and canine afterlives (at least) were one and the same.

Our relationships with dogs are psychologically, physiologically, and socially encoded. Both parties have adapted to experience a rush of oxytocin (the 'cuddle hormone') during dog–human bonding sessions, making our relationships both socially and *somatically* encoded. Dogs are particularly attuned to the multisensory body language of *humans*, including any facial expressions and olfactory signals that could reflect potential illness or distress.[18] The fact that dogs frequently intervene in our moments of need with displays of affection, concern, and closeness is a further indication that they not only recognize when a human is in pain, but also understand and attempt to attend to us in those situations.

While we'll find no accounts of the true 'first dog' in any ancient record, it should come as no surprise that dogs have important roles in many of our oldest mythic traditions. In many parts of the world, they were characterized as psychopomps and guardians of the underworld,[19] like Kérberos, the three-headed hound of Hades (and his Indic correlate, Śárvana), or the Welsh Cŵn Annwn. Dogs have long served as important intermediaries between human and non-human beings, but such 'otherworldly' associations weren't always so fearsome. In ancient Greece, dogs could often be found in the healing temples, nursing patients' wounds and offering therapeutic support. In traditions where unseen beings were still perceived as the chief drivers of disease, a dog's capacity to reach beyond the human world was seen as a tremendous magical (and thus also *medical*) asset.

We learned a great deal about the world from our relationship with dogs, not only because of the dogs themselves, but also because they represented an undeniable *other* with whom we somehow managed to enter into a meaningful relationship. By the time dogs and humans decided to share their lives, human 'religion' was primarily built upon an animistic worldview with shamanic modes of engagement. It's possible that tribal shamans were themselves responsible for navigating our early interactions with wolves, as 'making friends' with a notorious predator was surely perceived as a profound spiritual feat.

While such spiritual and otherworldly connotations have somewhat diminished over time, dogs still play very important roles in our 'human' society. They monitor our physical health, support us emotionally, assist us with accessibility, and sniff out dangerous weapons. A dog was even the very first earthling to venture into outer space. It's not difficult to see that dogs still, after all these years, stand at the threshold of the more-than-human world.

In the 20th century, trailblazing researchers like Jane Goodall helped to demonstrate to the world that non-human animals were emotional, conscious, sensory beings with the capacity to experience pleasure and pain – *living* beings much like ourselves. Naturally, this raised some important and concerning questions regarding the ethical treatment of animals, aptly laid out in ethicist Peter Singer's 1975 classic, *Animal Liberation*.

Our Vegetal Cousins

When we talk about *nature*, however, we need to realize that we are largely talking about *plants*. To the human mind, animals live *in* nature and humans *visit* nature, but plants are more or less *synonymous with* nature. This inseparability is, of course, the true condition of *all life* as we know it, but it's decidedly more obvious in those who seem to mostly stay put. It has been all too easy for us to see plants as part of a backdrop rather than *living entities* in and of themselves. They are, in fact, our distant cousins. It's worth stopping for a moment and allowing yourself to be fully gobsmacked by the recognition that you, your cat, and the flowers in your garden are all *genetically* related.

Over the past 20 years, thanks to advances in novel fields like *plant neurobiology*, many of our presuppositions about the conditions necessary for the emergence of intelligence in nature have been comprehensively challenged. We've begun to acknowledge that plants are complex *information-processing organisms* with sensory and 'cognitive' faculties that are not entirely dissimilar to those found in the animal kingdom. Even without a brain, plants make active decisions about how to nourish themselves, how to allocate molecules within their bodies, when to reproduce or generate new organs (like flowers), how to protect themselves against external attacks, and what messages they should send to their neighbors via chemical signals.[20] To make these decisions skillfully, they must take many things into consideration, including the weather, the ambient temperature, the moisture level, the availability of nutrients, the presence and identities of insects, and a great many other things in their ever-changing world.[21] Like many of us, plants also rely on 'memories' of their own past experiences to make educated decisions about how to engage in the world.

Plant neurobiology is a somewhat controversial discipline, though this is rapidly changing. Most lingering grumblings stem from the use of the term 'neurobiology' for organisms that simply don't possess *neurons*. Neurons are special cells that animals use to facilitate cellular communication across their multicellular bodies and it's long been assumed that neurological circuitry is indispensable for complex cognition and 'behavior.'[22] But the discoveries of plant neurobiology have comprehensively challenged this assumption. While the use of neurons is clearly the way that *animals* facilitate intercellular communication, non-animals have clearly found their own ways to do so, and our full understanding of these processes, and their relationships with 'consciousness' or 'awareness', remain in their infancy.[23]

Plant roots are a key area of interest for plant neurobiologists, particularly the *meristematic* tissue found in their tips and between their 'vascular' xylem and phloem tissue layers.[24] Rather than possessing a single colossal brain, plants may possess thousands of 'brain units,' all interconnected by vascular strands that transmit signalling molecules like *auxin* hormones and electrical impulses along a distinctly 'neuronal system.'[25] Plants have a vast chemical vocabulary, which they use to send messages both between cells and to other organisms.[26] While our understanding of the mechanics of plant behavior is only now beginning to clarify, one thing is certain: Plants are *not* unconscious automata – they are aware, intelligent, and somatically embodied *beings* with whom we have a great deal in common.

In most modern conversations regarding the environment, plants are spoken of in terms of their utility or abstract ecological importance rather than concern for their welfare. Even when we bemoan the loss of insects or forests, we almost always focus on the potential effects that this will have on *our* crops or air quality, rather than acknowledging the loss of life as its own tragedy. While this approach may seem quite rationally self-preserving, it fundamentally stifles our capacity to effect meaningful change, because it inhibits our ability to empathetically relate to the experiences of those who are most immediately and disproportionately affected by climate desecration.

Plants are not set pieces in our human drama, they are characters with distinct experiences and needs. Like us, they are capable of affection, care, and suffering. While we can (and do) dismiss the moral importance of plant experiences, to pretend that they don't exist is a very dangerous form of disregard. Our cold, utilitarian approach to mitigating climate change has so far been largely ineffectual. To bring real change, we need to learn to *see* plants as sovereign, intrinsically valuable, *sentient* beings – not mere tools for our own human flourishing. Our welfare is deeply dependent upon the welfare of plants, and the best way for us to make *sense* of this is by reminding ourselves that they are *persons*. They have just as much of a right to clean water, healthy soil, and a viable atmosphere as the rest of us, and this needs to be remembered when we consider the ethical dimensions of climate change.

The Life of Plants

What is the natural state of these beings? While our kinship is indisputable, many of the 'survival solutions' employed by plants appear to be the precise *opposite* of the approaches taken by animals. As Stefano Mancuso states:

> *Animals move, plants do not; animals are fast, plants are slow; animals consume, plants produce; animals make CO_2, plants use CO_2. But the most decisive contrast is also the least known: the difference between concentration and diffusion. Any function that in animals is concentrated in specialised organs is spread throughout the entire body of plants. This is a fundamental structural reason for why plants appear so different from us.*[27]

Different, but still *responsive*: *Mimosa pudica*, commonly known as 'the sensitive plant,' is an excellent representative of both the brilliance and *motility* of the plant kingdom, largely because they move on a timescale that can easily be perceived by humans. Any external physical disturbance causes them to quickly fold their leaves, to the shock and awe of onlookers.

There are various theories to explain the mechanics of this process, but the underlying *behavioral* framework is significantly more puzzling. Scientific research into *Mimosa pudica* began in the late 18th century, when French botanist René Desfontaines ordered a student to take eight potted plants on a carriage ride through the cobblestoned streets of Paris. As expected, the plants immediately closed their leaves at the first rattle of the carriage and remained indisposed for a while thereafter. Slowly they'd reopen, then close again at the next major disturbance. But as the journey drew on, the pupil noticed that the plants stopped reacting to the jostles. He scribbled in his study notes, perhaps not realizing the significance of his words, 'The plants are getting used to it.'[28]

Despite the radical implications of this observation, 'behavioral' research into the sensitive plant faded into obscurity until the 21st century, when evolutionary ecologist Dr Monica Gagliano, inspired by Desfontaines's account, decided to design her own experiment. Fifty-six potted plants were rigged to a machine that dropped them from a height of 15 centimeters every five seconds, with 60 drops per session. Startled by the movement, the plants predictably folded their leaves just as they had done on the streets of Paris over 200 years prior. But after a few regular drops, some of the plants stopped folding their leaves, suggesting a degree of habituation. They were 'getting used to it.' Or was this merely a result of exhaustion, and would the plants resume their instinctual behavior after a period of rest? To test this, the researchers disturbed them with a wholly different sensory experience, shaking them from side to side, and found that they somehow mustered the strength to fold their leaves in their usual fashion.[29]

When the researchers returned to test the plants after a week, they were amazed to find that the leaves remained open, unfazed by the now predictable drop, suggesting that they had somehow retained the lessons of the past experiment. The researchers repeated the test after a month – more than enough time for the plants to recharge – and the results remained the same. The plants had learned that the stimulus was non-threatening, and had augmented their behavior on this basis.

Many of Gagliano's peers balked at her use of terminology like 'behavior' and 'learning,' insisting that these terms were reserved for animals with neurological systems. But with increasing evidence supporting the behavioral sophistication of plants, we are forced to accept that cells other than neurons can facilitate intelligent and coordinated behavior.[30] Intelligence appears to be a fundamental feature of organic life itself, rather than a complex evolution unique to just a few species. Resistance within the scientific community is no longer an issue of insufficient data, but of unjustifiable theoretical bias.

In his 2022 book *Planta Sapiens*, philosopher Paco Calvo outlines yet another remarkable trait of *Mimosa pudica*. By simply placing a plant in a jar with a cotton pad soaked in veterinary anesthetic, he found that it was possible to 'put a plant to sleep.' After an hour of exposure, they lost all responsiveness to external stimuli and their leaves were devoid of any trace of motility until the anesthesia wore off.[31] Calvo has also performed this experiment with other plants, like the Venus flytrap (*Dionaea muscipula*), and found they likewise remain motionless while under the drug's influence. He has further demonstrated that the electrical signals produced by plants in their normal 'waking' state are completely halted under sedation – and this isn't limited to highly mobile plants. He writes:

> *It's not just mimosa leaves or Venus flytraps that lose their dramatic abilities under anaesthetic. All plants will stop whatever they were doing when under the influence, whether that be turning their leaves, bending their stems or photosynthesising. Seeds will even halt their germination. In short, anaesthetic causes plants to stop responding to the environment in all the ways they usually do.*[32]

Naturally, if plants can be made *unconscious*, this would suggest that they have a *conscious* state. That they are affected by *chemicals* is hardly surprising, since plant bodies are every bit as reliant upon chemicals as ours. In some ways, plants act as their own compounding pharmacies, producing special chemical 'drugs' tailored to their individual needs. If a plant is wounded or stressed, it can produce its *own* anesthetic (like ethylene) to 'take the edge off,' while other chemical responses can help to activate its support network and bolster its biological defenses.[33]

Plant-Sense

Another particularly mind-blowing example of the multisensory intelligence of plants is found in the Boquila plant. Native to the temperate rainforests of Chile and Argentina, *Boquila trifoliolata* is a rather nondescript jungle vine at first glance. Their small white blooms are far from extraordinary, and they remain rather modest in size. But in 2013, botanist Ernesto Gianoli stumbled upon something wholly extraordinary. While wandering through the southern Chilean forest, he came across a common shrub with which he was already well acquainted. But upon closer inspection, he found that some of the leaves looked a bit different from their standard form. He realized that the distinctive leaves did not, in fact, belong to the familiar shrub at all, but to a neighboring Boquila specimen *in disguise*.[34]

Boquila can, quite effectively, alter their physical appearance to resemble their vegetal neighbors – a form of mimicry that usually occurs only on a macro scale over many generations (presumably via natural selection), so plants can mask themselves from predators. However, Gianoli found that the highly responsive mimicry of the Boquila plant could respond to highly complex situations in 'real time,' with one single specimen contorting its leaves to resemble *multiple different neighbors at once, even in the span of just a few inches*. This even occurred in cases where there was no physical contact between the Boquila and the plant they were mimicking, leading to the question of *how*, exactly, the plant *knows* what its neighbors look like in the first place?[35] As mind-boggling as it might sound, Boquila likely knows simply because it can *see*.

Similarities between the lens-like shapes on the epidermis of plants and the primitive proto-eyes (*ocelli*) of many invertebrates first caught the attention of scientists over a century ago. These structures seem to provide plants with a sense of *vision*, allowing them to experience and respond to visual phenomena outside themselves.[36] Of course, the sensory perception of *light* (the basis of sight itself) is wholly integral to plants' evolutionary needs, given their dependence upon the light of the sun for photosynthesis. Even a simple beech tree, in order to avoid a premature growth spurt during an unseasonably warm spring, waits until there are at least 13 hours of daylight before beginning its yearly growth cycle.[37] This is, at least in part, based on a sensory perception of light. But if sight is indeed what facilitates the Boquila's special talent for impersonation, then these 'simple' visionary mechanisms are far more complex than we currently realize.

Plant Communities

Much like humans, plants form communities, communicating with one another through various pathways, including electric and chemical signals. They are linguistic experts in this regard, with a massive vocabulary that enables them to stay up to date on current affairs and reach out to others in times of need.

Both plants and animals can pick up on these chemical signals, though to varying degrees. While a dog has 300 million olfactory receptors in their nose, a human only has around 6 million. We can pick up on relatively strong smells, but this is nothing compared to the massive amount of sensory data that can be captured by many of our animal and plant relatives.[38]

Even a simple bean plant is a proper linguistic savant when it comes to chemical communication. When they are attacked by a hungry insect like a spider mite, they analyze the insect's saliva through a sensory process that we might relate to as 'taste.' Once they have determined the precise identity of the assailant (even down to the specific *subspecies*), they release a bespoke pheromone precisely formulated to attract the natural predators of the mite in question, broadcasting that message into the 'airwaves' through its leaf stomata. The predators pick up on these signals and are drawn to the plants, relieving them of their burden by eating the spider mites.[39] Between the sensory intake of the mite's saliva and the plant's production of a chemical signal, there's an obvious process of *analysis* based on previously acquired knowledge. Plants actually receive lessons from their elders, who pass along their wisdom using this same chemical language.[40]

Plants are sensitive to sound as well. Researchers have determined that the roots of grain sprouts 'crackle' at a consistently low frequency of 220 Hertz, but have been shocked to find that when other plants are exposed to this crackling sound, they reliably orient their tips toward its source.[41] Another study observed plants quickly increasing their nectar production (by an average of 20 per cent) within minutes after detecting the sound of a bee buzzing through their flower petals.[42] These behaviors should come as no surprise, given that plants have evolved to live in a noisy world with lots of important data transmitted via sound waves. But we are so alienated from this reality that we still have the audacity to ask if a falling tree makes a sound if no one is around to hear it.

In recent years, scientists like Dr Suzanne Simard have helped to introduce the world to the wonders of *mycorrhizal fungi*.[43] While not plants themselves, these enterprising beings work *with* plants to create complex forest *intranets*, connecting plants through their roots and facilitating communication and trade between a diverse array of organisms. More than 90 per cent of all plant species rely on this symbiotic partnership for their survival, making it a core component of what it means to be a *plant*.

We think of plants as 'rooted' beings, but roots were actually a late addition to the evolutionary toolkit. By the time plants started growing their own, they had been depending on root-like fungi for over 50 million years.[44] These thread-like mycelial 'landlines' can extend for miles in the soil beneath our feet,[45] enabling plants to transmit chemical signals, nutrients, and even healing compounds across long distances. If an individual in a tree community is sick, for example, their neighbors will send them resources to correct any nutritional deficiencies and support their recovery. Trees *care* for the wellbeing of their neighbors, even those that belong to entirely different species. This may all sound a bit fantastical, but it is in fact a function of the natural empathic magic of our planet. In undisturbed plant communities like old-growth forests, the oldest and most well-established trees are the ones who maintain the most mycelial connections, being either directly or indirectly *physically* connected to every other tree in the network. These 'Mother Trees,' as Suzanne Simard calls them, act like information and resource 'hubs,' sharing excess nitrogen and carbon with younger and weaker seedlings and boosting their capacity for survival.[46]

Plants have adapted to pick up on many different forms of sensory data, and to communicate intentionally with one another to form collaborative societies. But the communities that form are, in many ways, more than just the sum of their biological parts. Through unions such as these an entirely *new entity* is formed – a kind of compound organism, sometimes referred to as a *holobiont*.[47] We are technically holobionts ourselves, with more than half of the cells in our body belonging to *non-human* organisms like bacteria.[48]

Every tree in a forest can be thought of as a discrete 'person,' but the forest itself *also* comprises a special kind of complex 'person,' complete with an embodied, dynamically integrated, and sensitive corporeal form. This has often been posed as a rational scientific basis for 'nature spirit' motifs like a *god of the forest* or *river goddess*, and in fact *all* self-organizing natural systems could reasonably be personified in a similar fashion. This sense of personification can, if approached skillfully, effectively compel us to relate to such organisms and natural environments with not only respect but also *empathy*, thus compelling us to naturally limit our negative impact on their quality of life.

Magical Mushrooms

Without their fungal comrades, early plants likely would have never made their way out of the water 500 million years ago. Not only did mycorrhizal fungi serve as their very 'roots' for tens of millions of years, but even when they learned to produce their *own* roots, the fungi continued to make themselves indispensable.[49] Today, not only do the mycelia facilitate communication and trade between plants, they also ward off bacterial invaders, protecting the sensitive microbiome of the plant community and helping all life on Earth to thrive.[50]

But fungi aren't simply plant bodyguards and broadband providers. Like animals, they can't simply produce their own nourishment by sunbathing, so they rely on other organisms to sustain themselves, and, like all other forms of life, they actively communicate and collaborate with other beings to maintain a balanced environment and increase their odds of survival.

While it may come as a surprise, fungi are actually genetically closer to animals than they are to plants, but their uniqueness has earned them their own special kingdom in the tree of life. Without them, our world would look phenomenally different, and we almost certainly wouldn't be a part of it. Fungi are inside us and all around us, with nearly every stratum of our environment impacted in some way by fungal involvement.

A recent study found that mushrooms communicate via electrical signals in a manner that is not that dissimilar to human language, with strings of distinct 'words' and 'sentences' forming a vocabulary of at least 50 distinct patterns.[51] This study was conducted using only a few select species of mushroom (including the famed 'caterpillar fungus,' *Cordyceps militaris*, used in traditional Asian medicine), and is certainly only the tip of the iceberg when it comes to this form of fungal communication.

We may think of mushrooms as charming little toadstools in the forest, but they are the fruiting bodies of a fungus – a mere reproductive 'organ' used for spore production, often extraneously, since most fungi can release spores without producing mushrooms at all.

While considerably less romantic than mushrooms, *yeasts* are another kind of fungus without which humanity would look dramatically different. They're perhaps most famous for their capacity to transform sugar into alcohol, and their ability to facilitate the rising of dough for baking bread. Both of these innovations radically changed human societies, and many scholars believe that it was beer and/or bread that drove us to establish fixed agrarian societies in the first place.

Interestingly, medieval European alchemists like Paracelsus maintained that this alcohol, known as *philosophical Mercury*, was the common vitalizing *spirit* inherent in all life, particularly vegetal life, an idea that we still invoke every time we refer to alcohol as a 'spirit.' Any plant, if covered with water and left to ferment, will produce at least a little alcohol. While other components of a plant, like volatile essential oils and other phytochemicals, are often unique and distinct between species, the alcohol is always essentially the same.[52] Due to a natural abundance of yeast, even animal bodies ultimately ferment during decomposition, leaving behind traces of 'spirit.'

Of course, it would be remiss to talk about fungi without bringing up the most mysterious members of their family – *magic mushrooms*. Humankind has been experimenting with psychoactive substances since prehistoric times. Cave paintings dating back as far as 9,000 years support the theory that Neolithic humans were intentionally imbibing entheogenic mushrooms, though it's possible that we've been tripping on such substances for tens or even hundreds of thousands of years.

While entheogenic substances are illegal in most parts of the world – a lamentable effect of the draconian 'war on drugs' – modern research has shown that compounds like psilocybin can actually have profound healing effects. Serotonergic psychedelics like psilocybin, LSD, and DMT have a direct effect on the 5-HT2A receptors found throughout the body, which tend to go into overdrive in cases of depression. It appears that 'drugs' such as these disrupt and rewire these neural processes and facilitate greater overall integration and connectivity in the brain's network. In a study published in *Nature Medicine* in April 2022, researchers at UC San Francisco and Imperial College London found that psilocybin proved remarkably efficacious for treatment-resistant depression, disrupting patterns of negative thinking by allowing the brain to become more fluid and flexible.[53]

The capacity for psychedelic mushrooms to form healthier neural connections within our brains nicely reflects the role that mycorrhizal fungi play in the plant world. Most of the planet's natural grasslands are in fact specifically dependent upon the presence of psilocybe mushrooms for their ecological health.[54] Many beings depend on this symbiotic relationship, with fungal hyphae affixing themselves to plant roots and facilitating the exchange of nutrients and information across an entire ecosystem. But these are, of course, no ordinary mushrooms, and their effect on their community is no less profound than it is in an animal brain. The same serotonergic compounds that cause humans to 'trip' are diffused throughout the vegetal environment that the fungi sustain, facilitating increased intake of sensory data, enhanced communication between plants, and the capacity to adapt and innovate in response to external disruptions.[55]

Just as serotonergic psychedelics 'rewire' our brains to create a more globally integrated neurological system, they similarly compel plants to sprout new roots and expand the horizons of their own 'neurological' activity.[56] This is one of the reasons why psilocybe shrooms like *Psilocybe cubensis*, or our native liberty cap, *P. semilanceata*, are so frequently found on disturbed ground, especially land that has been overgrazed by cows and other livestock. This is, for the mushrooms, an ideal environment, with plenty of herbivorous dung and grass roots to keep them busy and fed for a long time. But there are also distinct benefits for the plants, and thus also for the animals who depend on the plants, and the many forms of life that depend on healthy soil.

Unfortunately, while evidence of the natural virtues of entheogenic medicines has been piling up for years, it has had a negligible effect on public policy concerning psychedelic substances themselves. If they are recognized, the commodification and medicalization of these 'drugs' will undoubtedly lead primarily to the production of synthetic pharmaceuticals, rather than a widespread acceptance of natural entheogens. Such things are naturally best engaged with under close guidance from trained experts, but little attention is paid to the many generations of expertise that can be found in indigenous societies that have worked with many of these substances for thousands of years, or to the agency of the plants and fungi themselves.

Mushrooms are obviously not the only mind-altering substances that we like to imbibe. We regularly ingest and smoke a wide array of plants, fungi, and even animal products to create a shift in our perceptual awareness. Some of these substances are relatively gentle, like tea leaves and coffee beans, while others, like *Salvia divinorum* and the peyote cactus, can blast us into what can feel like entirely different realities. But mushrooms are unique for many reasons. Ethnobotanist and famed psychonaut Terence McKenna argued that the use of psychedelic mushrooms may have had a direct role in the quantum cognitive leaps that occurred in our evolution, and while this theory remains quite speculative, it's certain that entheogens can be readily found in all corners of the world, and that humans are naturally inquisitive. It's certainly more than likely that we dabbled with psychedelics from time to time, and that any chance encounters that we *did* have with these substances would have left a rather strong impression.

According to oral traditions in South America, shamans have used psychedelic preparations like *ayahuasca* (made from the *Banisteriopsis caapi* vine) to commune with the plants, animals, and 'spirits' of the jungle for countless generations. These relationships have been maintained to this day, allowing indigenous healers to learn about the medicinal powers of plants *directly from the plants themselves.*

It's important for us to remember that mushrooms and *Banisteriopsis caapi* vines are not 'drugs' or 'resources,' but the actual physical bodies of living, sentient beings who can act as *teachers,* guiding us through the dynamism of the spirited world and the recesses of our own psyche into a deeper sense of *integration* with the world around us. It's not essential to adopt such teachers when traversing a spiritual path, but consulting them certainly shouldn't be stigmatized. There are risks that come with psychedelic use, just as there are risks associated with following a human teacher, but both have the potential to provide us with new and important information, some of which can radically expand our understanding of reality. It's important that we treat such non-human teachers with the same degree of reverence and respect that we would extend to our most beloved gurus.

Living in a Microbial World

Animals, plants, fungal organisms... What other non-human beings are unseen in our 'backdrop' of nature? Long before there *were* animals, plants, or fungal organisms, there were microbes. They are so ancient and so pervasively diffused that it's nearly impossible to truly comprehend just how many of them exist. There are more bacterial *species* on our planet than there are stars in our galaxy – even just a single teaspoon of healthy soil can easily contain up to a billion bacteria and many yards of mycorrhizal filaments.[57] Some scientists have estimated that there are over a *trillion* different *species* of microbe on Earth, against which macro-organisms like animals positively pale in comparison.[58]

In 1972, historian and geographer Alfred Crosby published a monumental book called *The Columbian Exchange*, which explored the transfer of plants, animals, and pathogenic microbes between the 'Old World' of Eurasia and the 'New World' of the Americas.[59] Beginning in the 15th century, Native Americans were involuntarily exposed to a perilous onslaught of foreign infections for which they had never had any opportunity to develop an immunity. As a result, countless indigenous lives were lost to diseases like smallpox, whooping cough, malaria, scarlet fever, and yellow fever as a result of European colonization.[60] But while the Columbian exchange was certainly one of the most disastrous epidemiological catastrophes in human history, it was far from an isolated incident. For more than 10,000 years, human societies have been gravely impacted by the threat of epidemic infections – a threat which clearly looms large even today.

While our modern understanding of microbial infections arose through the establishment of germ theory in the 19th century, it's important to recognize that this medical 'discovery' was not without precedents. The idea that tiny unseen organisms cause infectious diseases can be traced back at least 1,000 years to the works of pre-modern scientists like Ibn Sina (980–1037) and Girolamo Fracastoro (c.1476–1553). In medieval Europe, this theory was not only *known* to medical scientists but also hotly debated, as it ran counter to the prevailing Galenic notion that 'bad air' (*miasma*) was the true cause of pandemic infections. This Galenic model dominated much of Eurasian medical orthodoxy for around 1,500 years, but alternative theories regarding infectious diseases were also quite prevalent. In some cases, infectious diseases were conceptualized as a pathogenic attack from unseen 'spirits' who, although invisible, could be indirectly observed and placated through medical and ritual interventions. But to most of the educated medical practitioners of Europe, the notion that invisible beings could cause disease was deemed to be fundamentally superstitious – that is, until the latter half of the 17th century.

With advances in microscopic technologies, we quickly discovered that our entire world was positively teeming with 'unseen' entities in the form of microbial life. We now know our own human bodies contain more bacterial cells than human cells, with most of us hosting something to the tune of 39 trillion non-human microbes.[61]

Microbes are at once biological entities and truly *unseen* beings, generally invisible to the naked eye *and* routinely excluded from conversations about sentience and 'being-ness.' But, much like their more 'advanced' vegetal, fungal, and animal cousins, they also demonstrate the capacity to sense and respond to their environment, communicate with one another, and make decisions based on their lived experience.

Does this make bacteria sentient beings? Many scientists dismiss the possibility of microbial intelligence on principle, but the same has been true for fungi, plants, and even many non-human animals for thousands of years. As long as theorists remain fixated upon limiting and incongruent anthropocentric models, we will remain ignorant of the full range of awareness manifest in the organic world.

Viruses: Agents of Possession

When Spanish flu broke out in 1918, it was generally assumed that it was just another bacterial infection for which a suitable treatment could promptly be established. Doctors had only just heard of 'viruses,' and they were still very poorly understood. The modern term *virus* comes from the Latin *vīrus*, cognate with the Sanskrit *viṣa*, both of which mean 'poison.' While this is a rather charged moniker, we can be forgiven for perceiving viruses as evil toxins. They have caused some of humanity's most harrowing pandemics, including the Spanish flu, HIV, and

Covid-19, and even though we can now isolate and observe them through sophisticated technologies, there is still a great deal we don't know about them.

For starters, there is no scientific consensus on whether or not viruses are 'alive.' They are sometimes described as being 'at the edge of life,' lacking the basic cellular structures typical of living organisms and relying entirely on external beings for reproduction.[62] But if they are not 'alive' in a conventional sense, they still operate in a manner that is quite characteristically 'behavioral,' making a compelling case for their existence as non-biological *beings*.

Like bacteria, viruses are phenomenally pervasive. If we were to line up all of the Earth's nanoscopic viruses from end to end, they would extend all the way from Earth to the Camelopardalis constellation *100 million light years away*.[63] There are more viruses on Earth than all other life-forms combined, making these 'unseen beings' the most pervasive 'entities' in the known universe. They challenge our understanding of many things, not least of which is the origins of life itself.

There are numerous theories of how viruses emerged. Some scientists say they formed from stray bits of genetic material that learned to move between cells. Others think they descended from once-autonomous organisms, but over time they chose to embrace a *parasitic* way of life. Some have even suggested that they were *our* progenitors, and that all of us are descended from some virus-like ancestor.[64] Because they're made up of mere strands of proteins that unfortunately leave no trace in the archeological record, it's impossible to study a 'virus fossil.' But fortunately, their traces can be found in the genomes of other living beings. Researchers have detected traces of viruses in a wide range of organic genetic material, even in our own human genome – around 8 per cent of our own DNA (100,000 pieces) is viral.[65]

Scientists are still learning how our viral genes impact us. Oxford virologist Aris Katzourakis notes that, 'It's not an either–or – are these things good or bad? It's a lot more complicated than that.'[66] Some of these viral influences may protect us from diseases, while others may put us at higher risk. Many are forms of retroviruses, which can infect reproductive cells and be 'inherited' as endogenous stowaways. Over time, retroviruses mutate and evolve, losing their infectious potency. Our bodies become accustomed to their presence, and adjust accordingly, and it's been argued that a breakdown in this state of equilibrium is a major factor in the development of some tumors.[67] But these viruses have been at it for at least 450 million years, even impacting our distant ocean-dwelling forebears. They have been winding their way around our biological family tree since 'the beginning.'

The emergence of germ theory helped us realize that microscopic organisms were involved in many disease processes, but there were always those who cautioned against forming an overly simplistic view of their role in the disease process. As George Bernard Shaw once wrote, 'The characteristic microbe of a disease may be a symptom instead of a cause.'[68] Epidemiology is profoundly more complex than 'one germ, one disease.' Pandemics like Covid-19 arise due to an array of numerous environmental, social, and institutional factors, including deforestation, the wildlife trade, agriculture, travel, urbanization, and issues surrounding access to healthcare. We would never have been exposed to viruses like SARS if their bat hosts hadn't been forced into contact, quite unnaturally, with *Homo sapiens*.

While they may not seem like conservationists, viruses do not seek to destroy their own ecosystems. Once they've adapted to a particular biological host, their best chance of evolutionary success is to keep them alive and kicking for as long as possible. While we believe that Ebola is derived from fruit bats, we've never actually found a live Ebola virus in a bat specimen. What we *do* find are antibodies, which purge the virus from the bat's system once it has had *just* enough time to reproduce and spread to a neighboring bat. In this way, viruses can circulate through animal populations *ad nauseum*, 'feeding' off their 'prey' without actually harming them.[69]

But just as a farmer might shoot a wolf who gets too close to his flock of sheep, viruses can become volatile when confronted with rival predators. Once a secure relationship has been established, viruses will 'protect' their flock from invaders by making the strangers sick. A good example of this is the simian herpes-B virus (*Herpesvirus saimiri*). For its natural hosts, the virus causes absolutely no issues, but for foreign monkey species with no natural immunity, it produces a deadly and fast-acting lymphatic cancer.[70] For the host monkeys, these viruses might even be characterized as 'protectors' – and in the right contexts, viruses *can* be tremendously helpful allies. We've even managed to recruit some of them for therapeutic purposes, as in the use of bacteriophages to treat bacterial infections.[71] But when it comes to the 'environment,' viruses force us to reckon with our own hubris.

Nature Will Not Be a Passive Victim

There are some places we simply *should not tread* – and if we choose to do so anyway, we must be ready to face the consequences.

As Mark Honigsbaum writes in *The Pandemic Century*, 'Unless and until we take account of the ecological, immunological, and behavioral factors that govern the emergence and spread of novel pathogens, our knowledge of such microbes and their connection to disease is bound to be partial and incomplete.'[72] While a deep understanding of viruses themselves is essential for the mitigation of pandemics like Covid-19, we *must* also pay attention to the non-viral 'causes' of such crises, especially our broken relationship with nature. Pandemics are, in many ways, symptomatic of broken relationships – between humans and humans, and between humans and non-humans. In a very direct way, the *microbiome* and *virosphere* force us to see *all* of nature as interconnected, sensitive, and suffused with awareness.

We have overlooked this because, while neural structures certainly play a vastly important role in the cognitive processes of most *animals*, plants, fungi, amoebas, and bacteria have managed to get by quite well without them. But, while there are certainly profound differences in the ways that a vertebrate, a slime mold, and a mushroom experience 'reality,' we have no reason to believe that the *experiencing* itself is confined to only *some* forms of life. The evidence points strongly to the contrary – subjective awareness is all around us, even in the most unlikely places. In *all* of us, the quest for survival pushes us to pursue the things that will benefit us and to avoid the things that will harm us. This in itself is an expression of 'sentience,' and it is staggeringly pervasive in the natural world.

What we *do* with this knowledge is of the utmost importance. If genuinely metabolized, it has the capacity to radically alter our approach to life on this planet.

When viewed without the filter of anthropocentrism, it is in fact quite easy to see our own reflection in the beings and systems of the natural world. This is, quite simply, because we *are* beings of the natural world. Humans *are* nature. But nature is not *only* human. It manifests in myriad forms, including every skyscraper, every song, and every thought. We cannot escape nature. We cannot master it. For millennia, many of us have simply been asking the wrong questions. Now, rather than trying to find a way to dominate nature and assert our agency over it, we should instead explore ways of *relating* with our fellow nature-beings ethically and meaningfully in order to support our collective flourishing.

3

BALANCE AND IMBALANCE

The Animistic State

For most of our species' history, we have been deeply and consciously invested in the affairs of the more-than-human world. We didn't think of ourselves as being existentially separate from 'nature,' but rather an integral part of it. To be *in relationship* with non-human beings – friend, foe, food, or otherwise – was a deeply important part of what it always meant to be 'human.'

This *relational* approach to the living world is what most scholars now call 'animism.' But while we render it as an '-ism,' animism is neither a religion nor a system of 'belief,' but a paradigm of more-than-human relationship. It can even be thought of as an *evolutionary adaptation*, given its ubiquity in global hunter-gatherer societies and its profoundly ancient origins, likely predating even the rise of complex language and religion (including gods, ancestor veneration, and conceptions of an afterlife) in hominids.[1] One might argue that, philosophically speaking, animism is our 'natural state.'

Animism stands very much in opposition to anthropocentrism, being the radical acknowledgment that human and non-human *persons* pervade, and even *comprise*, the so-called 'natural world.' When viewed with this kind of orientation, the 'exploitation of nature' stops being a simple matter of appropriating resources and instead can be identified as systemic violence perpetrated against myriad living beings. While a certain degree of 'violence' is an intrinsic part of being an animal on Planet Earth, animism pushes us to *not turn away from it* or brush it under the rug to make it more palatable. If we can remember that there are real *living beings* on the receiving end of our exploitative practices, we'll naturally be more compelled to take their needs and experiences into account. It is *this* empathic experience, and not bureaucratic rules and regulations, that will ultimately inspire the most meaningful shift in our human behaviors.

On the Alleged 'Primitiveness' of Animism

Animism first entered academic discourse in the 19th century, with the work of Sir Edward Tylor, who defined it as 'the general doctrine of souls and other spiritual beings in general,' based on the 'primitive' notion that vitality and agency pervade both animate and inanimate phenomena.'[2] For Tylor and his peers, this belief system represented a core rational *error* in human thought. He argued that animism was an essential precursor to religion because it taught us to conceive of the existence of *spirits* and *supernatural* entities. But in a 'rational' world in which the non-human domain is populated merely by biological automata, all such constructs were seen as utter fictions. But, as Bruno Latour writes:

... animation is the essential phenomenon; and deanimation is the superficial, auxiliary, polemical, and often defensive phenomenon. One of the great enigmas of Western history is not that 'there are still people naïve enough to believe in animism,' but that many people still hold the rather naïve belief in a supposedly deanimated 'material world'.[3]

Modern research has largely confirmed Tylor's theory of the primality of animism,[4] but many of his further conclusions have proven to be both unsubstantiated and dangerous.[5] While Tylor himself avoided making any sweeping judgments regarding the intellect of indigenous animists,[6] allegations of cultural *primitiveness* quickly became a political tool to bolster support for the subjugation and colonization of indigenous societies. Colonial apologists argued that anyone who was unable to recognize the difference between a 'person' and a 'thing' was fundamentally *unevolved*, and thus incapable of managing their own social and political affairs. Animism was, at best, an exotic novelty of naïve primitive societies, but at worst it was seen as a dangerous delusion to be forcefully 'civilized.'

To claim new lands and resources for their waning monarchies, European colonists decimated and dispossessed countless indigenous communities around the world – killing, enslaving, and raping hundreds of millions of human beings in a process that is far too infrequently identified as a holocaust. The destruction was never limited to beings and bodies. Innumerable lineages of generational wisdom, environmental knowledge, stories, rituals, and ways of *being* in the world were severed. Those who weren't killed were stripped of their land, their traditions, and their more-than-human communities through forced 'conversion' and assimilation.

Anthropocentrism and a disregard for nature didn't come to dominate the world through organic *diffusion*, but through propaganda, violence, forced indoctrination, and colonization. Of course, once 'the West' realized that *disenchantment* wasn't the fantastic liberating force that it was expected to be, efforts at cultural erasure shifted to mere exploitation and appropriation, and soon we began treating 'exotic' foreign spiritual practices as commodities to fill the gaps in our own mutilated cultural traditions.

The Dawn of Humanity

It's important to remember that humans weren't just plopped onto the Earth out of nowhere in the distant past. It's believed that *Homo sapiens* emerged in Africa around 300,000 years ago, but our ancestors were already very much here, in myriad forms, and had been for over four *billion* years. Animism is, at its core, a dynamic recognition of this fact.

Around 100,000 years ago, a bunch of *Homo sapiens* decided to leave their homeland and venture northward out of Africa, just as past hominids like *Homo erectus*, and even some stray *H. sapiens*, had done in ages past. It's likely that they migrated due, in part, to environmental factors, but for reasons unknown most ultimately decided to return to Africa. Then, around 40,000 years later, a second mass migration was attempted, also likely instigated by environmental stresses, and this time the early humans successfully established themselves in Eurasia, ultimately expanding out to the Americas and Oceania.[7]

By this time in our prehistory, human proto-religion had already come to involve some distinctly 'religious' features, like a belief in an afterlife and the emergence of 'shamans' tasked with facilitating interactions between human and non-human forces.[8] In hunter-gatherer societies, the shaman's role became deeply important – shamans led negotiations with the spirits of the land, encouraged social bonding through communal ritual, and facilitated healing through their medical knowledge and engagements with non-human forces. It's likely that they had a pivotal role in humanity's early migration processes, offering healing and emotional support in times of great uncertainty in a wild and foreign environment.

One of their tasks may have been establishing relations with other hominids, since there were numerous hominid species already living in Afro-Eurasia. Neanderthals had been in Europe for over 300,000 years before Homo sapiens decided to settle in – and far from being some brutish cavemen, these other humans were much like us in many respects. They made art, buried their dead, and likely engaged in ritual activities of their own.[9] Many early human species may have used some form of verbal communication, and *Homo erectus* may have even built seafaring vessels to settle on the islands of Crete and Flores.[10]

We're not entirely sure what happened to these other hominid species, though our own ancestors likely played a role in their disappearance. But it's clear we lived alongside them for some time, and procreated with them on occasion, which itself may have had a detrimental effect on their own populations.[11] It's possible that we treated them as kin, at least for a time, and who knows what the 'human' world would look like had they never disappeared. Even now some of these early hominids actually live on in our own bodies, with most modern humans containing at least some genetic admixture from species like the Neanderthals, Denisovans, *H. habilis*, and *H. erectus*.[12]

Early *Homo sapiens* flourished, both in Africa and beyond, due to many factors, particularly a talent for community-building. Like many animals, we tend to form close-knit social relationships, and, like all animals, we also have relatively fixed upper limits for manageable community sizes. Such limits are usually dependent upon how many direct relationships a single individual can comfortably maintain. When existential 'strangers' begin to seep into the mix, conflict arises, often leading to discord and disunity.

As humans, we're usually limited to around 150 meaningful relationships at any one time, which also happens to be the general population limit for human communities over much of our early prehistory.[13] What was it that allowed us to go beyond these limitations to form societies of millions (and even billions) of people? It all comes down to the power of myth.[14]

In this sense, *myth* refers simply to *invented realities* or 'fictions' that exist solely in the space of shared imagination. Notions like nations, the economy, corporations, and religions are all ultimately socially encoded myths, upheld by rituals, stories, and our collective 'belief' in their existence.[15] In many cases, myths nourish and sustain us, while in others they can lead to delusion and self-destruction.

Myths of all kinds have played critical roles in our human development, providing us with a sense of order in dizzyingly vast and complex communities. Even though we can never establish trust-based relationships with a billion strangers, a shared myth can go a long way in establishing a sense of fundamental 'kinship.'

Around 33,000 years ago, glaciation gradually increased across much of the Northern Hemisphere, dramatically altering the hospitability of our new domains. By 24,000 years ago, glaciers engulfed much of Europe and North America, and even Tanzania's Mt Kilimanjaro was completely sheathed in ice. Our survival was more precarious than ever, and we were forced to rise to this challenge with collaboration and ingenuity. We developed an array of new tools and weapons during this time, from replaceable spearheads to sewing needles, and our hunting and foraging practices became far more calculated and organized. Our relationships with other beings changed drastically as well, including our budding alliance with the wolves. To make sense of these changes and challenges, we relied heavily on the power of myth.[16]

A 'Spirited' Environment

Our conceptualization of *anim-ism* tends to deal mainly with a belief in *souls* and *spirits*, or *anima*. Naturally, concepts like these are wrapped up in a great deal of cultural baggage, and their modern meanings have quite little relevance to traditional animistic worldviews. Over the past century, scholars like Irving Hallowell, Nurit Bird-David, Terri Smith, and Graham Harvey have helped academics and seekers alike to embrace a more nuanced approach to animism – sometimes termed *new animism* – which treats it not as a belief system but as a paradigm of relationship with the more-than-human world. Harvey writes:

> *Animists are people who recognize that the world is full of persons, only some of whom are human, and that life is always lived in relationship with other. Animism is lived out in various ways that are all about learning to act respectfully (carefully and constructively) towards and among other persons. Persons are beings, rather than objects, who are animated and social towards others (even if they are*

not always sociable). Animism… is more accurately understood as being concerned with learning how to be a good person in respectful relationships with other persons.[17]

Animism is, importantly, *not* a religion. While it can certainly serve as a *basis* for any number of philosophical or religious ideas, it can just as easily remain a domain of open-ended natural observation. It has no fixed dogma, nor is it particularly 'supernatural,' and it emerged ages before humans ever conceived of an afterlife. It wasn't some 'primordial error' that propelled us into delusion, but rather a reflection of our natural state. It pushes us to relate with the diverse array of beings that surround us in a nuanced and empathic way – an invaluable skill for surviving in a world replete with competing wills and perspectives. Animism didn't trigger our ontological disease; it was actually a mark of good health.

From an animistic perspective, *all* of nature can be seen as potentially cognizant and aware. The extent of this can vary significantly from animist to animist – while some perceive every rock and grain of sand as 'alive,' others take a somewhat more conservative approach to identifying sentient beings. While some traditions established particular 'gods' or 'spirits' within these dynamics, deities are not historically fundamental to the animistic worldview. Special relationships with certain plants, animals, or mountain spirits might naturally arise through engagement, but the idea that beings exist solely to act as tribal gods or tutelary deities for humans is somewhat antithetical to the animistic ethic. Humans are spirited (i.e. aware) beings living in a spirited (i.e. aware) universe, and all of us are in relationship with one another.

The Hidden Folk

For many of us in the 'west,' *animism* evokes images of tribesmen wandering through the jungle in grass skirts. We think of it as a somewhat whimsical function of 'indigenousness.' But it is not the exclusive provenance of uncontacted tribes or remote hunter-gatherers. Even in Europe, there are countless stories, rituals, celebrations, and folk traditions that have their roots in animism. And while it's vitally important that we acknowledge and respect the unique cultural traditions of indigenous peoples and the ways in which they *diverge* from our Euro-American societies, it is equally important that we don't reinforce the perceptual divide between 'us' and 'them' by bypassing these important points of intersection.

In Iceland, for example, it's estimated that around half of the population, even today, retains a traditional belief that elves (*álfar*) and *hidden folk* (*húldufolk*) inhabit the natural landscape. Such beliefs are a well-accepted part of Icelandic life and have been for many centuries, though in recent years they have received increased attention with regard to environmentalism. In 2013, protestors successfully halted the construction of a new road near Reykjavík due to the threat it posed to a local elf habitat,[18] and this is far from a one-off occurrence for this small island nation. It turns out that elves carry a non-negligible amount of clout in Icelandic law and politics, even in the 21st century.

Elf lore in Iceland stretches back to its early pagan days, though written evidence only begins to emerge in the Christian 11th century. In the Christian era, some came to believe that elves were the offspring of Lilith, or even the forsaken children of Adam and Eve. According to the latter legend, which is quite popular in Iceland, Eve hid their unwashed children from God out of embarrassment, causing him to make them invisible in return. But in reality, Iceland is one of the least religious countries in Europe, and elf beliefs are often entirely outside paradigms of Christian theology.

Elves are, however, closely connected to the Icelandic landscape. In the 2019 documentary *The Seer and the Unseen*, Icelandic grandmother Ragga Jónsdóttir fights to protect a lava field that she perceives to be inhabited by a colony of elves. As a self-described 'seer' with the capacity to communicate with the hidden folk, Ragga has a clear, deep, and obviously sincere emotional reaction to the razing of the lava field. Many Icelanders' relationship with the elves, and thus with the landscape itself, is on a deep emotional level.

While beliefs in 'hidden folk' were once common to every society on Earth, over time they became twisted into anthropocentric tropes, and were often abandoned outright in the establishment of monotheism. Reliable sources regarding the pre-Christian beliefs of European societies are often hard to come by, simply because writing was often introduced as a function of Christianization. Works like *Beowulf*, the *Mabinogion*, and the Norse *Eddas* were all ultimately written down by Christians, and at best offer only a small slice of the ancestral myths associated with 'pagan' traditions. In the Mediterranean world, where people were writing down their ideas and beliefs for many centuries before Christianization, 'heretical' works were frequently destroyed or otherwise lost to time, leaving us with precious few written fragments of pre-Christian Europe. Regardless, to think that the older traditions wholly died out would be quite a mistake.

Elves were relegated to children's stories and nursery rhymes, for example, but Anglo-Saxons held on to their own *ælf* beliefs for some centuries after Christianization, and we can even find accounts of Christian clergymen intentionally seeking engagements with elves, faëries, and other hidden folk. East of Loch Lomond in Scotland, in a village called Aberfoyle, Reverend Robert Kirk, who served as the local minister until 1692, became known as the 'Fairy Minister' due to his obsession with Doon Hill, an enchanted hill behind his parish. He was a renowned Gaelic scholar and theologian who contributed to the first Scots translation of the Bible, but he also claimed to possess a 'second sight' that allowed him to observe and interact with the hill's bustling community of faëries, and the latter part of his life was largely dedicated to this 'research'.[19]

Kirk's death was itself steeped in mystery. He died on a walk on his beloved hill, and many came to believe that he had been taken captive by the faëries and made to serve as an intermediary between their world and the world of humans. To this day, it's held that he is embodied in a lone Scots pine growing at the top of the hill, where visitors still leave prayer ribbons and offerings in hopes that the arboreal reverend will pass along a good word to the local faërie folk.[20]

While the Enlightenment effectively stamped out many such traditions, the 19th century saw a massive revival of faërie and elf lore across Europe, with folklorists seeking to re-enchant a world that had become 'increasingly dominated by reason and industrialism'.[21] Most of these movements fizzled out by the mid-20th century, with some notable exceptions to which we'll return later.

The Evolution of the Unknown

Even though we now know a great deal about the material structure and movements of the universe, countless mysteries remain. One of our most beloved modern myths is the phenomenon of *unidentified flying objects*, or UFOs, now termed *unidentified aerial phenomena* (UAPs). These are, in many ways, a 'spirit' paradigm for the technological age. If you were to witness a strange glowing orb hovering above marshland, you might well conclude that you were looking at an *extra-terrestrial* spaceship, whereas an Anglo-Saxon farmer might assume it to be the magical display of an elf or nature spirit.

Most *supernatural* experiences are sensorily vague, often involving wisps of light, disembodied voices, glowing orbs, flying disks, and shadowy figures. With some notable exceptions, our experiences alone rarely provide the necessary data to explain the nature of the phenomena, and so our interpretation is conditioned by our own preconceptual biases. In the 21st century, the biases of *technocratic anthropocentrism* have left us with an acceptable 'supernatural' framework centered around humanoid extra-terrestrials conducting science experiments in spaceships, but this still remains a story built around *ourselves*.

The unexplained is undeniably alluring, but we should be wary of sacrificing our sense of earthly enchantment for yet another paradigm of transcendent escapism. The Earth is plenty enchanting, with many mysteries yet to be uncovered, and if we want to find signs of 'intelligent non-human life,' all we need to do is look around.

Embracing Non-Human Personhood

In 2018, the Nonhuman Rights Project filed a petition for a common law *writ of habeas corpus* to demand that Happy, a 40-year-old elephant held at the Bronx Zoo, be released from her captivity and transferred to a suitable wildlife sanctuary. They argued that Happy deserved legal protections as a *person* because she was an 'extraordinarily intelligent and autonomous being who possesses advanced analytic abilities akin to human beings.'[22] Unfortunately, the highest New York court struck down the petition in June 2022, deeming the case a 'slippery slope' for dealing with matters of alleged animal sentience.[23] Despite an outpouring of support from around the world, Happy remains a captive *object*.

The notion of *non-human personhood* is often used in a legal context to establish rights and protections for certain non-human animals. It's a tremendously useful concept, but when *speciesism* seeps in, things do tend to become slippery. The Nonhuman Rights Project is itself only focused on *animal* personhood specifically, mainly the rights of 'chimpanzees and elephants,' though 'other potential clients include orangutans, gorillas, bonobos, dolphins, and whales.'[24] Why these animals in particular? According to the Nonhuman Rights Project, it all comes down to scientific evidence of 'self-awareness and autonomy,' defined as follows: 'Self-awareness is the capacity to *recognize yourself as an individual separate from the environment and other individuals*. Autonomy is the capacity to make choices about how to spend your days and live your life.'[25]

Despite the project's noble intentions, this definition of self-awareness is not only anthropocentric but also deeply *ethnocentric*, and not unrelated to the colonial arguments that were levied against animism. Our recognition of our separateness from our environment is, by these standards, a *precondition* for the possession of basic rights – quite the Catch-22 when the future of humanity is dependent on our ability to recognize ourselves as *a part of the environment*. Furthermore, to think that only a handful of *animals* make agentive decisions in their lives is quite mind-boggling – all organisms, including humans, make decisions based on genetic, biological, and environmental factors.

Even a beloved elephant who's managed to jump through these anthropocentric hoops to defend her moral worth legally remains a *non-person*. But there's reason to be hopeful. In some places, special trees and rivers have been granted some degree of legal personhood, like the Ganges and Yamuna rivers of India. Such distinctions have been (arguably symbolically) bestowed to combat pollution, but if efforts like these can rise above mere symbolism, they *could* potentially revolutionize our discourse surrounding environmental ethics. If protecting nature is only ever touted as a means to an end for protecting *humanity*, then we will most certainly fall short of our goals. We need to come to embrace *all* plants, animals, and natural ecosystems as *persons* in their own right, and earnestly confront the ethical questions that result, rather than dismiss non-humans entirely out of fear of 'slippery slopes.'

Unfortunately, we are only now beginning to embrace more robust moral paradigms for the treatment of *animals*, and the consideration of *plant rights* remains quite the stretch goal. Yet plants remain some of the most vulnerable beings on our planet, and our dependence upon them makes *us* vulnerable as well. If we could actually respond to their mistreatment as a moral issue, then we might be able to forge more robust legal mechanisms for protecting nature and decelerating the loss of species.[26]

But we must be wary of becoming overly fixated on the notion of 'biological organisms.' A 'being' isn't always limited to the body of one organism. A forest, for example, is both a community *and* a kind of 'person.' As with other ecosystems, its component beings are deeply integrated into the whole, making the welfare of the entire 'being' integral to the welfare of each individual.

There *are* growing movements centered around a more *eco-centric* approach. While they don't deal with 'personhood' taxonomies, groups like the Community Environmental Legal Defense Fund (CELDF), a US non-profit organization, work with partners in Asia, Africa, Oceania, and the Americas to 'craft and adopt new laws that change the status of natural *communities* and *ecosystems* from being regarded as property under the law to being recognized as rights-bearing entities.'[27] Within such a model, all beings and natural communities are extended a fundamental right to 'exist, flourish, and naturally evolve.'[28]

In 2008, CELDF helped Ecuador to become the first modern nation to adopt this approach on a national constitutional level, using legal measures to enshrine values that had long been a part of the indigenous animistic traditions. Its ratified constitution extends fundamental and inalienable rights to nature, personified as *Pachamama*, including the rights to *exist*, to be *protected from harm*, and to be *restored*. If (and when) nature's rights are violated, there are systems and procedures in place to ensure that the offenders are held accountable, that the environment itself is restored, and that the rights of nature are not undermined by private interests. Importantly, since nature is unlikely to file its own legal case when a crime has been committed, human citizens can petition on behalf of a violated ecosystem to ensure that justice is served.[29]

But such projects cannot be limited to developing nations. In the UK, the late Scottish barrister Polly Higgins led a decade-long campaign to establish *ecocide* – 'unlawful or wanton acts committed with knowledge that there is a substantial likelihood of severe and either widespread or long-term damage to the environment being caused by those acts'[30] – an international crime, proposing an amendment to the Rome statute of the International Criminal Court (ICC) that would make it punishable by international law. In 2015, she wrote:

> *The rules of our world are laws, and they can be changed. Laws can restrict or they can enable. What matters is what they serve. Many of the laws in our world serve property – they are based on ownership. But imagine a law that has a higher moral authority... a law that puts people and planet first. Imagine a law that starts from 'first do no harm', that stops this dangerous game and takes us to a place of safety....*[31]

In all that we do, we should strive to *see* non-human beings as agents and 'persons,' but this doesn't mean we must treat all such 'persons' exactly the same. It's self-evident that we all need and desire different things. But these are nuances and details that must be addressed *once we have accepted* that non-humans are agentive 'beings' who must be included in our sphere of moral concern. Everyone has different needs, and some needs will necessarily trump the needs of others in some situations, but that doesn't mean that only *human* needs matter, or that only humans have a basic right to exist. The world was *not* created for us, and we are *not* the only *persons* who live here.

Acknowledging Our Mistakes

While a basic acknowledgment of the vitality and agency of non-human persons was once commonplace worldwide, this 'primitive barbarism' is now seen to have been gradually 'civilized' through the establishment of city-states, bureaucracy, and (European) religion.

Thomas Hobbes's *Leviathan* (1651) was instrumental in establishing this narrative in the western psyche, and went on to form the basis of many modern European political systems.[32] Hobbes based his approach on what he described as the 'natural condition of mankind,' i.e. the state of humanity *before* the rise of civilization. In his view, our primal state was one of violence and chaos – 'a war of all against all.'[33] Early humans were nothing more than terrified, impoverished, brutish, uncultured, short-lived ignoramuses, wholly incapable of cooperation.[34]

Modern trends in 'big picture' history tend to either follow this Hobbesian model or its inverse, exemplified in Jean-Jacques Rousseau's *Discourse on the Origin and the Foundation of Inequality Among Mankind* (1754). Rousseau's alternate version is far more idealistic, imagining primitive hunter-gatherers to have been egalitarian tribespeople living in a state of relative harmony and only rarely engaging in violence for self-preservation.[35] According to Rousseau, it was the establishment of civilization itself, and more specifically the concept of private property, that caused us to fall from this naturally peaceful state. While Hobbes imagined humanity to be perpetually progressing, Rousseau imagined us to be perpetually *regressing*.

But, as Graeber and Wengrow demonstrate in *The Dawn of Everything*, both approaches to history are fundamentally myths that grossly over-simplify the complex evolution of human societies.[36] We have always been highly experimental and mutable in our methods of social organization, with no discernible *arc of history* that has been in any way preordained. Our story is still very much being written, and even now it remains a 'choose your own ending' adventure.

Unfortunately, Graeber and Wengrow still present ours as a fundamentally *human* story, as if it is human decision-making alone that matters in the grand scheme of things. This, too, is a contrived philosophical notion – and one that severely limits our capacity to imagine a new future. We need to relinquish our quest for dominance. We need to learn to take a step back and *listen* – to other persons, and other kinds of knowledge. It's clear that the European colonists were *wrong* about non-human beings and the inanimacy of nature. We have long undermined indigenous worldviews based on the assumption that they are nonsensical, but really the opposite is true.

This isn't just a matter of 'respecting other cultures,' as if the acknowledgment of non-human agency were some whimsical cultural practice like circle dancing. The annihilation of indigenous traditions robbed humanity of *knowledge*, not just *culture*, and indeed the specific forms of knowledge that are most needed in our present crisis.

So what now? We cannot, in good conscience, just say 'Whoops!' and steal what remains of the world's indigenous knowledge. We can and *should* certainly learn from whatever wisdom their stewards are willing to impart, but to really make a difference we need to reflect critically on our *own* traditions and worldviews, and, more importantly, our relationships with our own environments. This is a pathway open to us all. We don't need to don fancy robes or take on a new name to deepen our relationship with nature. As naturalist John Burroughs writes, 'There are no heretics in Nature's church; all are believers, all are communicants.'[37]

Perhaps – just perhaps – there will someday be justice for the plants, animals, rivers, oceans, and other unseen beings that share our world. But there must also be justice for the humans who have unwittingly been caught up in this project of domination. It's not that nature lacked allies in the human world, or that humanity is inescapably primed for domination and exploitation. But those who maintained the closest relationships with the Earth were often the first to be sacrificed on the altar of 'progress.' This remains the case today. Indigenous peoples around the world are often on the frontlines of climate activism, putting their lives on the line to defend the agency of the more-than-human world.

According to many scholars, this era of European colonization hurled us across the threshold of the Anthropocene. Besides the toll it took on human societies, landscapes, animals, plants, and ecosystems were all forever changed in the process. But was this era of violence truly the 'root cause' of our climate crisis? Probably not. But with a somewhat clearer understanding of how we've diverged from a state of health, we may find it easier to recognize where things went wrong in the first place. This will provide us with some important perspective regarding our condition and its treatment.

PART II

CAUSES AND CONDITIONS

4

NATURAL HIERARCHIES

Philosophical Roots of the Great Divide

Now we've familiarized ourselves with the 'natural' state of things, the *etiology* and *pathogenesis* of our condition must be earnestly assessed if we're to make any progress in treating our ecological crisis. Of course, our relationship with nature evolved gradually over time, with ever-shifting social, cultural, and geopolitical dynamics perennially ratifying the ways that we have perceived and engaged with our environment.

While many have justifiably pointed to our religious institutions as a primary trigger for our anthropocentric beliefs, faith alone cannot be blamed for our ecological disregard. Many of our most secular western institutions, including the hard sciences themselves, have perpetuated anthropocentrism as a 'rational' creed, and the formalization of these prejudices predates our reigning religious institutions by many hundreds of years.

Descartes, the 17th-century father of modern philosophy, often takes a great deal of blame as the key driver of anthropocentrism.

He concretized the concept of a fundamental divide between *body* and *mind*, famously elevating the mental phenomena of *humans* above the material domain, and characterized all forms of non-human life as mere mechanistic *automata*, devoid of any subjective awareness, or 'soul.'

But by the time of the *Cartesian split*, the 'roots of disregard' had already been creeping into our collective psyche for thousands of years.[1] The *unseeing* of non-human beings was a project shaped by many hands over time. Descartes was, like many Renaissance and Enlightenment thinkers, deeply influenced by the 'great minds' of the classical world, and while he certainly challenged many of their ideas (often to align more cleanly with Christian dogma), his dismissal of non-human sentience was an uncritical perpetuation of long-standing delusions dating back to ancient Greece.

Cooperation, Cultivation, and Control

Looking deeper into our prehistory, it's clear that our relationships with other species took a dramatic turn during the Agricultural Revolution.

Like many such 'revolutions,' the development of agriculture was a long and complicated process. Across the world, human societies had been experimenting with cultivating plants and migrating with flocks of animals for tens of thousands of years. We had been 'in relationship' with other beings in one way or another well before the rise of agrarian and pastoral societies. The average hunter-gatherer had *significantly* more practical knowledge about their environment than most of us do today, specifically because they relied upon *nature* for survival. There's a great deal of evidence that the very brains of our foraging ancestors were *larger*, and potentially more powerful, than ours today.[2]

As ancient humans made their way out of the Last Glacial Maximum, which reached its peak around 20,000 years ago, life became gradually easier with each passing generation. In places like western Asia, harsh conditions gradually gave way to a warmer and wetter climate, leading many groups of people, like the Natufians, to establish semi-sedentary communities in the lush woodlands along the Mediterranean and rivers like the Euphrates.[3] By 15,000 years ago, hunter-gatherers in the Levant were experimenting with cultivating grains and baking bread, and by 13,000 years ago they were brewing beer in modern-day Israel.[4] These technological developments were naturally accompanied by changes in beliefs and social structures. Burials in this time became highly complex, with around a quarter of corpses adorned with elaborate headdresses and jewellery made from animal bones and seashells, presumably as a marker of wealth or social rank. Remains were also periodically disinterred and reinterred, likely as a part of post-mortem ritual proceedings.[5] Wherever it was that the dead were believed to be going, they were certainly going in style.

But this new semi-sedentary lifestyle wouldn't last forever, and things took a dramatic turn in what's known as the *Younger Dryas Period*, a 1,000-year spell of rapid and intense climate change likely triggered by a cosmic impact 12,850 years ago.[6] Conditions became dramatically colder and dryer across much of the Northern Hemisphere during this time, and within 500 years things looked much as they had done in the Last Glacial Maximum.[7] The Natufians were forced to abandon their bread and beer and return, like many others, to a far more difficult subsistence lifestyle.

When this mini-Ice Age finally began to let up, communities in western Asia and beyond started domesticating and cultivating plants and animals in earnest, some focusing on the cultivation of cereals and others specializing in pastoral livestock rearing. These developments led to yet another explosion of cultural innovations and brand-new social experiments centered more exclusively around agriculture.[8]

Around a century after the end of the Younger Dryas Period, construction began of one of the world's oldest and most remarkable prehistoric wonders, Göbekli Tepe in Turkey. This remarkable megalithic complex is believed by some to be the world's first temple, being used from around 9500 to 8000BCE, though its utility clearly extended far beyond our modern understanding of a temple. It was one of *many* megalithic sites built in the area, all of which were constructed, used for ritual feasts, and then ceremonially buried.[9] But these incredible 'temples' weren't the centerpieces of some kind of Neolithic city – Göbekli Tepe was built and used by nomadic hunter-gatherers. This rather shocked historians, who had long assumed that such an architectural feat could only be accomplished by a complex agrarian society, and that such monumental 'religious' structures could only arise *out of* established civilizations. But as archeologist Klaus Schmidt put it, 'First the temple, then the city.'[10]

Evidence shows that at least one of our varieties of domestic wheat came from a site about 18 miles away from Göbekli Tepe, leading some historians to suggest that the temple complex itself sparked some of our more serious experimentations with an agrarian lifestyle.[11] Others disagree, viewing the complex as a kind of coordinated 'last stand' for the region's hunter-gatherers – a monument, as bizarre as it may sound, to the *'old* ways,' standing firm against a rapidly changing world.

But Göbekli Tepe was not a static monument – it was a living *ritual space*. Its construction and interment seem to have been a part of a ritual that began not long after the Younger Dryas incident. A team of researchers at the University of Edinburgh have argued that the art of Göbekli Tepe itself narrates a 'cometary encounter' leading to the environmental misery of this mini-Ice Age.[12]

While some disagree with the team's conclusions, it's quite interesting to think of Göbekli Tepe as a reaction to environmental changes, rather than some kind of spontaneous emergence of human creativity. These sites are perhaps not 'temples' at all, but the buried remains of massive ritual procedures that took place over thousands of years.

At this point in our story, it would be a mistake to think that such a ritual space belonged to the gods above. If anything, it was consecrated to the *underworld*, and to the enchanted domain of nature itself. Most of the beings depicted on the site's towering T-shaped pillars are distinctly *wild* animals, like scorpions, bears, spiders, and snakes – animals who have relatively little use as a source of sustenance, but who have deep spiritual importance to many cultures around the world. To the hunter-gatherers of the early Neolithic, it's clear that these beings were not mere resources or symbols, but dynamic and spirited *beings*.

But for many ancient peoples, the dramatic ups and downs of a chaotic shifting climate inspired comprehensive behavioral shifts. The Agricultural Revolution saw a massive array of domestication events across the world, allowing humans to control and manipulate sources of food rather than rely on nomadic foraging. Dogs, of course, helped facilitate this process, providing us with protection and even assistance with livestock. But over time, horses, oxen, sheep, cows, grains, chickens, and even cats also became integral parts of our human world, each adapting to our needs as food, workers, commodities, and pets. Though compared to dogs, and to a lesser extent horses, most of these relationships were far from reverential.

Horses themselves really shouldn't be overlooked, because our bond with our equine companions is itself wholly extraordinary. While wolves and humans are both apex predators, horses and humans are entirely different. Horses' bodies and brains evolved to protect them from a litany of ever-present dangers; humans' bodies and brains are pre-programmed for predatory behavior. The union of horse and human has been described as 'a neurobiological miracle,' with each animal functionally extending its neural network to include the other in a bi-directional feedback loop, facilitating highly complex and synchronized movement.[13]

Even though humans continued to treat horses as a source of food long after adopting them as steeds, it's evident that their domestication was, at least in part, a function of *relationship* and not mere exploitation. Horse-riding techniques are relatively universal, even among horses and riders that have been isolated from one another for millennia, and horse-riding is an inescapably collaborative endeavor – one that is rooted in mutual trust.

The benefits of agriculture and livestock domestication came with many world-altering consequences, and not just in our approach to animals and plants. Agriculture forced us into a painful confrontation with a whole new host of *unseen beings*, to whom we'll later return. It was also in large part due to the proliferation of agriculture that we gradually rose to prominence – in our own minds, at least – as the 'lords of the Earth.'

Epics of the Living Earth

Even in the Bronze and Iron Ages, most agricultural Eurasian societies were still *animistic* in many respects. In the *Iliad* and the *Odyssey*, two of our earliest works of western myth, the 'great gods' of legend are largely indistinguishable from natural forces. Poseidon, the 'blue-maned god who makes the islands tremble,' is *actually* physically embodied in the crashing waves of the sea, and the goddess of the dawn, with her 'spreading fingertips of rose,' can be sensorily experienced in the rays of an expanding sunrise.[14] But beyond these proto-divinities there is no shortage of 'lesser' spirits, like *dryads* and *nymphs*, in the hidden places of the world, often personified in the elements and living features of the landscape. David Abram describes Odysseus's world, and the broader environment of ancient Greece, as 'a land that is everywhere alive and awake, animated by a multitude of capricious but willful forces, at times vengeful and at other times tender, yet always in some sense responsive to human situations.'[15] Epics like the *Iliad* portray the forces of the natural world as *characters* in and of themselves, not passive background décor in a human drama.

We know very little about the historical Homer, and his authorship of the *Iliad* and the *Odyssey* is possibly itself a myth. Most contemporary scholars believe these two works to have been composed by different authors entirely sometime around the ninth to eighth centuries BCE, based on earlier traditions.[16] By the time these stories were written down and circulated as *texts* rather than living bardic traditions, they had already had creative inputs from a variety of sources, forming some of our oldest examples of recorded 'history' in Europe. Obviously, it would be inappropriate to read these texts uncritically as *historical* accounts, and the ancient Greeks were often quite self-conscious of this fact.[17] But myths such as these weren't intended to provide an objective chronicle of affairs, but rather an inspiring celebration of the heroes of yesteryear within a domain of more-than-human enchantment.

The celebration of past heroism was also a driving motivation for 'proper' Greek historians, like Herodotus (484–425BCE), the so-called 'father of history.' He recorded some of the first non-mythic chronicles in the Mediterranean world, his most famous work being his account of the Greco-Persian wars. Historians often view this transition from the myths of Homer to the histories of Herodotus as a key moment of evolution in perceptions of historical 'truth.' While Homer's epics were mythically rooted in an animistic and enchanted world, Herodotus began a pattern of historical demystification, balancing his curiosity about the past with an attention to evidence and reason. This ethic was more fully actualized in the later works of Thucydides, who largely established our dominant view of history as a process of rational inquiry, rather than a principally creative artform.[18]

Writing itself played a key role in this transition. Sometime around the 9th or 8th century BCE, the Greek world began to adopt a unifying *phonetic* alphabet derived from the script of the Phoenicians.[19] The adoption of a concise phonetic script, as opposed to a complex and expansive pictographic one, was a major step toward cultivating widespread community literacy, and by the end of the 4th century BCE anybody who knew the 24 letters of the Euclidean alphabet could effectively engage with textual materials across the expanding Greek world. But this didn't happen overnight, and it was only around the time of Herodotus and Plato that Athenian Greece could be considered an ostensibly *literate* culture.[20]

This transition to a new writing system was not welcomed by all in the Greek world, especially not by the professional bards, or *rhapsodes*, whose monopoly on oral histories made them the privileged knowledge-keepers of their societies.[21] But in the centuries following these developments, many novel works began to emerge that would lay the groundwork for a new class of knowledge-holders. Pre-Socratic philosophers like Thales of Miletus (7th to 6th century BCE) described a world that was still very much vital and alive, famously proclaiming that 'all things are full of gods,'[22] but by the 4th century BCE, the Greeks had already initiated a gradual departure from the magic of a living world.

Plato and the Soul

Plato is likely the most famous western philosopher in history, and his views have had an enduring influence on both religious and secular worldviews for thousands of years. He was a student of Socrates, himself the *prototypical* Greek philosopher. Rather than simply repeating the words of their forebears, Socrates encouraged his interlocutors to repeat, reformulate, and challenge their points of view from multiple angles, freeing themselves from the shackles of tradition in favor of a more engaged and critical mode of thinking.

Plato's works marked a turning point in ancient Greek perceptions of nature and *human reason*. There are still traces of animistic thought in his works, but his most impactful theories propelled the Mediterranean world unwaveringly toward anthropocentrism. In his *Timaeus*, we see the first conception of a *tripartite soul*: the first attempt at formally establishing a *natural order* of ensouled beings. This was an influential contribution to the changing Greek attitudes toward non-human beings.[23]

According to Plato, all living beings – humans, animals, and plants – are composed of the same primordial *stuff*, but out of this have come three types of *soul*: the *Appetitive Soul* is possessed by *all* beings (including plants), providing a basic *appetite* for food and sex; the *Spirited Soul* can only be found in humans and animals, producing instinctual emotional experiences like anger; the *Rational Soul*, the highest of the three, only ever arises in humans, and is considered to be most emblematically manifest in the freemen of Athens.

In Plato's view, those with the *Rational Soul* are the only individuals with genuine agency and moral value. As Val Plumwood writes in *Feminism and the Mastery of Nature*:

> *Platonic philosophy is organized around the hierarchical dualism of the sphere of reason over the sphere of nature, creating a fault line which runs through virtually every topic discussed, love, beauty, knowledge, art, education, ontology... In each of these cases the lower side is that associated with nature, the body and the realm of becoming, as well as of the feminine, and the higher with the realm of reason.*[24]

In effect, Plato's worldview was not strictly *anthropocentric*.[25] It was *rationality*, not *humanness*, that made an individual important – and while not all humans were deemed to be rational, all rational beings were necessarily human. But as his views spread beyond Athenian culture, the human-centeredness of his philosophy remained most pertinent. While we should be careful not to jump to the conclusion that Plato is *solely* responsible for all the world's ills, formal western arguments for androcentrism (i.e. male-centrism), anthropocentrism, and the backgrounding of nature can indeed be traced back to this 'great man' of Greek philosophy.

The Aristotelian Hierarchy

While Plato laid the groundwork for a new anthropocentric worldview, it was his student Aristotle (384–322BCE) whose contributions would have the most lasting impact, influencing our study of physics, esthetics, music, governmental politics, biology, economics, linguistics, and 'natural philosophy' well into the modern era.

In his formation of a comprehensive 'natural hierarchy,' Aristotle built upon Plato's tripartite soul to establish his own version:

1. *The Nutritive* or *Vegetative Soul*, bestowing the most basic capacity to grow, feed, and reproduce, is possessed by all living beings, including plant life.[26] Because all the faculties of this soul are wholly dependent upon a material body (if there's no body, there's nothing to grow or feed), Aristotle concluded that the Nutritive Soul must cease to exist upon death.

2. *The Animal Soul*, responsible for facilitating perception, desire, and locomotion, is possessed by all animals, including humans. It is also dependent upon the physical body and does not persist after death. It *does not manifest with subjective*

sensory or cognitive awareness, meaning that animals are fundamentally unconscious of their lived experience.

3. *The Rational Soul*, producing the lucid intellectual faculties of reason and subjective awareness, is a uniquely *human* soul. It is not confined to a particular feature or organ of our physical body, so Aristotle deems it to be immortal, but he clarifies that it is not the *personality* of an individual that continues after death, but the intrinsically divine *intellect*, which arises through dependence upon the other conditions of life.[27]

Aristotle taught that good moral conduct should be based on moderation and the cultivation of intellectual virtues like theoretical wisdom, science, and practical knowledge, which we should use to analyze the ethics of any given situation. On this basis, Aristotelian ethics *could* lead us to the rational conclusion that environmental destruction is a non-virtue rooted in the vice of pursuing excess. But this would only be the case if it negatively impacted human beings. If only animals and plants were negatively impacted, then Aristotle's existential hierarchy would undermine any necessity to change course. Humans are, fundamentally, the only subjects of moral consequence in the Aristotelian worldview. As he writes in his *Politics*:

> In like manner we may infer that, after the birth of animals, plants exist for their [animals'] sake, and that the other animals exist for the sake of man, the tame for use and food, the wild, if not all at least the greater part of them, for food, and for the provision of clothing and various instruments.[28]

This *hierarchy of use* is inextricably linked to the hierarchy of the tripartite soul, establishing a firm philosophical justification for the instrumentalization of animals and plants, who may be freely exploited for the desires of humankind.

Aristotle also perpetuates Plato's notion that some humans, like women and enslaved people, are metaphysically inferior and more akin to animals than the rational man. The enslavement of nature on 'rational' grounds ultimately helped to justify human enslavement, as well, so long as the enslaved person could be sufficiently *dehumanized* on a theoretical basis.[29]

Ancient Influencers

Aristotle wrote and published hundreds of books over the course of his life, although only about a third survive,[30] and his intellectual legacy throughout Eurasia was outstanding, especially in the development of the natural sciences.

There is a special kind of magic to be found in ancient texts. Reading the words of humans long past carries a thrill and sense of mystery, as if we're peering into a distant world. We might marvel at the humor or precocious wit of ancient authors or shake our heads in dismay at their mistakes. But for many, ancient texts remain absolute authorities, their mere existence legitimizing religious and cultural traditions around the world.

But for all the time we spend poring over ancient texts, far too little time is dedicated to the question of *why* these texts have survived in the first place. Most of what was recorded in the ancient world is now long lost. For Latin literature alone, it's estimated that around 99 per cent was destroyed in the early centuries of the Common Era, largely at the hands of religious zealots seeking to systematically scrub any trace of heresy from the written record.[31] What we're left with is simply what was allowed to persist. The fact that we can still read the works of Plato, or Aristotle, or the Christian Bible indicates that they were intentionally preserved, copied, and perpetuated over many generations of cultural change, and asking *why* can be wildly illuminating.

Theophrastus: The Better Botanist

When it comes to the backgrounding of *plants*, it should be noted that Aristotle's understanding of the botanical world was highly rudimentary and included some glaring errors. His own student Theophrastus (371–*c*.287BCE) was a far more gifted and passionate botanist, and it seems that Aristotle encouraged him to pursue the wonders of the plant world to fill in any gaps in his own natural philosophy. Theophrastus was critical of his predecessor's tendency to apply zoological concepts to the study of plants, demonstrating that they were fundamentally different from animals in most respects. But he also remarked on the distinct *cognitive* powers of plants, identifying them as complex autonomous and independent entities. 'The world of plants is manifold,' he reported, and certainly not observably inferior to animals.[32]

Just like his teacher Aristotle and his grand-teacher Plato before him, Theophrastus wrote his own treatise titled *De Anima*, 'On the Soul.' Unfortunately, this is one of the many ancient texts that have been lost to time. But in the scattered fragments and references that do exist, we can see that Theophrastus considered sensorial awareness to be wholly inseparable from the phenomenon of life itself – and as such, plants are regarded as distinctly *sentient* beings with agency, preferences, and the capacity to suffer.[33] Theophrastus argued against the alienation of plants from the domain of human morality. As Matthew Hall writes, 'Theophrastus understands cultivation to be a collaborative, mutualistic *relationship* between plants and humans. With this, [he] envisages a more respectful form of farming, in which the cultivator engages in a partnership based on respect for the awareness and autonomy of the cultivated.'[34]

While Theophrastus is still honored as the classical 'father of botany,' few of his works survived even in the centuries after his death, and his impact on later Greek and Roman thinkers was highly limited. Pliny the Elder, who wrote in the 1st century CE, does rely on him for his discussion of trees and plants, but he still adheres to the Aristotelian hierarchy of beings.[35] Thus the Aristotelian worldview endured as the most sophisticated and irrefutable basis for scientific investigations into nature.

But there was still significantly more nuance in the pre-Christian Mediterranean world. Pliny bemoans the exploitation of the Earth's resources by greedy men and speaks of a sacred and sensitive planet, albeit one that man rightly rules as the 'highest species in the order of creation,'[36] writing in 77CE:

> *We pursue all the lodes in the earth... and then are amazed that sometimes the earth gapes open or begins to tremble, unwilling to believe that this might be our holy parent's way of expressing her indignation... But what she has hidden and kept underground – those things that cannot be found immediately – destroy us and drive us to the depths. As a result, the mind boggles at the thought of the long-term effect of draining the earth's resources and the full impact of greed.*[37]

There's something remarkably strange about reading such words written nearly 2,000 years ago. These arguments are perhaps more relevant today than ever before, and could quite easily be applied to oil extraction or fracking. We now know much more about the long-term effects of extraction processes, but the mind certainly continues to boggle at the masochistic insatiability of greed.

From Disregard to Demonization

Christianity is a distinctly Greco-Roman religious movement in many respects, and as such is heir to many of the philosophical developments of the ancient Greek world. In addition to obvious Jewish and possible Zoroastrian influences, Christian rituals such as the Eucharist are clearly linked to the *mystery cults* that dominated Greco-Roman religion, and early Christians demonstrated a close affinity with Neoplatonic and Aristotelian worldviews. The standardization of Christian doctrine was an intentional and careful process drawing on many different sources to craft a definitive corpus for a future state religion.

But much of the classic Greco-Roman corpus was lost to European societies for many centuries, surviving only in the Middle East. Most of the Aristotelian corpus was only translated into Latin (from Arabic) in the 12th and 13th centuries.

Discovering that Aristotle was actually opposed to the concept of an immortal creator god, authorities in the Catholic Church promptly censured his works to avoid stirring dissent. But Thomas Aquinas (c.1225–74) sought to establish a more durable synthesis between Christianity and Aristotelianism, relying heavily on concepts of the *Rational Soul* to demonstrate both the existential superiority of humans and the notion that divine truths can be uncovered through a human process of *inquiry*.[38] So, it was largely their mutual *anthropocentric* grounds that allowed these two traditions to come together.

According to Aquinas, humans have no divine duty to act charitably or morally toward non-human beings, because they are fundamentally resources, not persons, having been produced specifically *by God* as *objects for human enjoyment*. He does caution against overindulging in animal abuse, but only because cruelty toward animals might lead to cruelty toward humans.[39] No act of violence against non-human beings is inherently sinful.

The Cartesian Split

While Plato, Aristotle, and later Abrahamic religious movements set the groundwork for our rejection of the non-human world, the Enlightenment era brought a new breed of philosophers, who sought to form new theories and philosophies that could propel Europeans into a brighter future.

By then, Christianity had effectively stifled any new technological or philosophical innovations in the West, and the perceived center of gravity for scientific progress had long shifted from the classical Mediterranean (if it was ever there to begin with) to Central and East Asia. Islamic theorists, in particular, had made great strides in astronomy, mathematics, and medicine, while Europe was stuck with fragmentary shreds of a long-gone Golden Age. But the 'discovery' of the Americas provided a new opportunity for fading European monarchies, in the form of 'new' and exploitable land, peoples, animals, and plants. It was desperate acts of colonization, not a linear trajectory of progress, that shifted the balance of Eurasian power – quite artificially – back into the West.

A little over a century after Christopher Columbus 'sailed the ocean blue,' a child was born in France whose views would remain a dominant intellectual force well into the modern era. His name was René Descartes.

Descartes (1596–1650) is often regarded as the father of modern philosophy, laying the foundations of modern rationalism with significant influences from both mathematics and his Catholic faith. He sought to balance a mechanistic view of the universe with Christian theology and Aristotelianism, proposing that everything in existence could be explained through the laws of mathematics, but that *the mechanics themselves* were placed in motion by the thrice-omni Christian God.[40] While he acknowledged that the human body was mechanical and 'natural,' he argued that the human *mind* was a special and distinct *non-physical substance* that bestowed both sentience and agency upon human beings.

To demonstrate his confidence in the soullessness of non-human beings, Descartes famously performed public vivisections on dogs and other animals, often to the shock and horror of onlookers. But as he and his students spread the dogs' legs, nailed their paws to wooden planks, and ruthlessly dissembled their still-living bodies, Descartes assured his audience that what might *seem* to be the screams of a creature in existential agony were *actually* just the pre-programmed responses of an unconscious automated machine.[41] Nothing to worry about, since animals don't have souls, and without a soul there can be no conscious 'experience.'

The Cartesian 'split' between mind and body has been picked apart from numerous angles for the past 400 years. We've gone well beyond Descartes' self-righteous musings in many aspects, but when it comes to our perceptions of non-human sentience, his utter disregard for the value of non-human life continues to echo in our modern world. While our standards of treatment for dogs and other 'pets' has largely improved, many remain convinced that an animal is essentially a 'machine made by the hands of God.'[42]

Later philosophers challenged Descartes' theories, however, particularly regarding the role of the *senses* in the acquisition of knowledge. While Descartes viewed the senses as ultimately misleading and unreliable,[43] John Locke (1632–1704) asserted that it was *only* through sensory experience that we could apprehend objective truth. But scholasticist perceptions of the natural world were not so easily relinquished. In his 1690 *Essay Concerning Human Understanding,* Locke neatly followed the Aristotelian model of the tripartite soul for his own rehashing of the 'Identity of Vegetables,' the 'Identity of Animals,' and the 'Identity of Man.'[44] Like Plato, Aristotle, and Descartes before him, he concluded that consciousness was an exclusively *human* domain, thus making humans the only holders of intrinsic 'natural rights.'

The Rise of Empiricism

While empiricists like Locke had a massive impact on the *philosophy* of western sciences, the greatest *scientific* breakthroughs of the 17th century emerged through the work of his contemporary Isaac Newton. Newton's laws of motion and universal gravitation were thoroughly paradigm-altering. He proposed that the motions of the cosmos could be understood through a basic set of physical laws, codified in the 'language' of mathematics, and that phenomena on Earth were governed by the same laws as astral bodies. There was, importantly, no need for divine agency in this model, and while Newton was nominally Christian, he was functionally a 'heretic' in many respects (though he didn't go out of his way to make this known).[45]

The Newtonian model of the universe ultimately paved the way for many great scientific breakthroughs, including in the realm of biology. In the 19th century, Charles Darwin applied it to creation itself, proposing that *all life* evolved through a similar set of mechanical, automated processes governed by natural laws, rather than the creative agency of an omnipotent God. However, even though he demonstrated, for the very first time, that plants and animals were actually *related*, this did little to dissolve the deeply rooted philosophical biases surrounding sentience.[46]

By the beginning of the 20th century, the Newtonian and Darwinian spirit had dislodged any rational sense of *divine involvement* in the workings of the universe. Even the most complicated phenomena could be understood by simply *reducing* them to their smallest parts. Reality was understood to be objective, reducible, and ultimately deterministic. But in the 1920s, the Newtonian model itself began to crumble with the discovery of quantum mechanics. Quantum theory severely challenged notions of a 'purely objective' physical reality, forcing the scientific model of the universe to shift once more.

In western thought, *objectivity* and *subjectivity* are seen to be diametrically opposed ideas – if there is no *objective* truth, then everything must, therefore, be *subjective*. We only allow ourselves these two options because we're still deeply invested in the fundamental split between subject/object and mind/matter.[47] In order to resolve this scientific identity crisis, a new paradigm of thought emerged in the 1930s, which remains the dominant view to the present day: logical empiricism.

Logical empiricism is founded upon two basic principles – *logic* and *empirical observation*. Propositions are devised based on *observations* and *experiments*, the results of which are translated into mathematics and used to establish logical *theories*. These theories are then used to predict further observations, which are either verified or refuted through further experimentation. This model is assumed to underlie all modern sciences, even though its practical application can vary widely.[48]

Of course, even assuming the model is followed faithfully, there is always the possibility that seemingly well-established theories will not be universally 'true.' We can look at a swan, see that it's white, and form a theory that all swans are white. We can test this theory by examining a large quantity of swans, and if all of them are white then we can presume that our theory is correct. But this would be a hasty conclusion, even with a data set of a million swans, because there are, in fact, non-white swans.[49] In this way, 'objective' empiricism is much better for showing us when we're wrong than it is for showing us when we're right, and it's generally accepted that scientific theories formed through this *inductive* process must remain *theories* (even if well substantiated), and not definite facts.

Yet, while scientists are often open to challenging *some* established theories, many theories are clung to with tremendous zeal, causing us to filter out 'anomalous' observations that run contrary to our existing model. Discussions of plant *behavior* are an excellent example of this. Despite years of research and piles of evidence, many critics still argue that studying the behavior of plants is fundamentally futile, simply because it's *theoretically* untenable.

Similar blockades have long plagued the integration of traditional medical therapies (like herbalism) into mainstream medicine – in many cases, criticism of a particular therapy is not due to its demonstrable efficaciousness (or lack thereof), but the fact that it does not fit into well-established theoretical paradigms.

It strikes me that, when it comes to science, we want to have our cake and eat it too. We want a scientific paradigm that's flexible, open to innovation, and capable of changing course based on evidence, but we're reluctant to actually acknowledge when the science tells us we're wrong. We struggle to see how many of our existing institutions are still rooted in obsolete models of reality. Whether it's trickle-down economics, conversion therapy, or the idea that livestock animals don't actually *experience* suffering, it seems that knowledge gained doesn't *always* lead to lessons learned, especially when money is involved. Revolutions require far more than new discoveries – they require a shift in worldview. And while philosophy and science have important parts to play, they are not the only forces that have informed the ways we view the world.

Our anthropocentrism, in particular, wasn't merely developed as a result of philosophical innovations. Rather, it is *religion* that has likely had the greatest impact on the spread of our disease.

GODS AND DEMONS

The Sanctification of Ecological Disregard

For most of our history, there was nothing inherently divine, diabolical, or even 'supernatural' about *unseen beings*. Much like humans, they were viewed as integral parts of our living world, often tied to the inner experiences of plants and landscapes – all equipped with their own desires, needs, and perspectives. As agents in a complex social world, they could also act as *aggressors* or *allies* in their relations with others, depending on the circumstances. The moral ambiguity and nuanced *personhood* of non-humans was once a universal basic assumption that most humans had about the nature of the universe. It wasn't until relatively recently that we began to augment our worldviews to better suit our anthropocentric delusions, and religion played no small role in this process. Over time, institutions emerged which sought to sanctify the centrality and dominance of humans in a world of *gods, demons,* and *resources.* This theological and religious evolution was an important spiritual progression in our existential disease.

The Old Gods

As we saw in Chapter 3, the spiritual lives of our hunter-gatherer ancestors were quite unlike what we think of today as 'religion.' Long before the emergence of holy books, temples, or even farms, the spiritual domain was often indiscernible from the normal activities of daily life, and life itself was deeply rooted in the living world. Animistic traditions pervaded every inhabited continent on Earth, transforming into innumerable regional variations of indigenous natural engagement.

Over time, conceptions of an afterlife emerged, along with paradigms of shamanic engagement, ancestor worship, and ultimately the creation of lofty and wilful 'gods.' Gods themselves were a relatively late addition to human religion, becoming increasingly prevalent with the rise of agricultural and pastoral lifestyles and the major social and cultural changes they inspired.

As cities and well-tilled fields began to replace the wilderness as our sphere of engagement, our spiritual needs also changed substantially. To ensure fertility and a bountiful harvest, agrarian societies often turned to divinities associated with the earth, rivers, and weather, who frequently manifested as maternal or otherwise feminine 'goddesses.' But pastoral societies, who specialized in livestock rearing, had very different spiritual needs. Many such groups relied on a largely masculine pantheon of powerful and often brutal gods, who were called on for protection and assistance during episodes of migration, conflict, and conquest.

It's important to recognize that, in the 21st century, the word 'god' can refer to many things. In most contexts, *God* with a capital 'G' refers to the thrice-omni *All-father* of so-called 'Abrahamic' religious traditions,[1] who has a very detailed backstory, absolute power and knowledge, and strong preferences regarding human behaviors, appearance, and belief. In some other contexts, when people talk about *God*, they're speaking about the ineffable 'source of all phenomena,' the mystical answer to all of life's riddles. In other contexts still, *gods* can refer to any number of powerful entities found in myths and religious traditions across the world, few of which were ever intended to be thought of as immortal creators or overlords. But the conflation of these three types of 'god,' both linguistically and theologically, has led to some massive misunderstandings.

But beyond a belief in gods (which not all religions share), the range of behaviors that we associate with 'religion' are largely natural and universal components of the human experience. We are naturally predisposed to engage with such constructs, presumably because of the distinct evolutionary advantages that they offer in terms of durable cooperation and group cohesion.[2] This proclivity likely predates even the emergence of modern *Homo sapiens*, with evidence of ritual behavior found among Neanderthal sites dating back 175,000 years,[3] and even being seen in other species like elephants.[4]

But the notion of 'religion' was itself a relatively late innovation in western thought, being put forth by European theorists in the 16th and 17th centuries as a universal classification for a wildly diverse array of global cultural traditions involving some combination of myth, ritual, mysticism, theology, and paradigms of morality.[5] Notably, most of the traditions that we identify as bona fide 'religions' are also literary in nature, with sacred texts of usually divine provenance. It's worth considering how such paradigms have impacted our perceptions of the so-called 'spiritual world.'

There is naturally little that we can *definitively* know about the specific beliefs and practices of prehistoric peoples, due mainly to the lack of a written record. Even the infamous myths of Old Norse religion, with gods like Odin and Thor, are only known to us through fragments of lore recorded by Christian writers centuries after the conversion of Iceland. But fortunately, we are not totally left in the dark. Many of our most important discoveries regarding the evolution of myth and religion have come from *philological* research.

The PIE World

Philology is the study of the development, evolution, and structure of, and relationships between languages. It approaches linguistics much in the same way that evolutionary biology approaches an organism, with each language comprising a single branch or leaf in a massive family tree. Early philologists proposed that, through comparative analysis, we could establish definitive 'genetic' relationships between even seemingly distant languages (like English and Sanskrit), and potentially identify a common ancestor. This hypothesis proved to be correct, and scholars have been able to map out the evolutionary lineages of all our major extant linguistic groups.

The Indo-European language family is the world's largest, containing around 450 languages, and Indo-European languages are today spoken by nearly 50 per cent of the world's population.[6]

This language family includes all Indo-Iranian, Germanic, Celtic, Greek, Italic, Balto-Slavic, Armenian, and Albanian tongues, meaning that all these linguistic lineages descend from a common ancestral language: Proto-Indo-European (PIE).[7]

But languages are never *just* languages. The words we use to describe the world can have a dramatic impact on the way we perceive reality. Languages are shaped by the stories we tell – and these stories are, in turn, shaped by our language. Both have a part to play in the establishment of *culture*, which is why we can rightly refer to 'cultural and linguistic *groups*' like the Proto-Indo-Europeans.

Proto-Indo-European is a *proto-language*, meaning that it is attested through linguistic comparison and reconstruction, not written evidence. It is the 'genetic' similarities between words, phrases, and semantic concepts that have allowed linguists to 'reconstruct' many components of the language used by PIE speakers 6,000 years ago.[8] Languages offer important insights into the worlds of their speakers, including cultural and historical developments. What was abandoned? What was adopted? When? How and why did meanings change over time?

When these philological processes are applied to *mythology*, we can start to unravel some of the most ancient stories told by our distant ancestors. This process is in no way limited to the Indo-European world, but given that 50 per cent of the global population lives in a cultural climate that has been significantly impacted by Indo-European languages, it's important to look at their legacies.

While the Agricultural Revolution significantly altered our approach to spirituality, the early Indo-European cultural sphere remained distinctly 'animistic' in many respects, albeit infused with an ethos of conquest and an expansive array of divinities. The speakers of Proto-Indo-European inhabited the Pontic-Caspian steppe, a culturally and genetically diverse region that witnessed many periods of multidirectional migration.[9] Like many of the steppe's other inhabitants, PIE speakers maintained a predominantly pastoral and nomadic lifestyle, utilizing horses and wheels to expand the geographic reach of their cultural and linguistic milieux (often with no small degree of force).

While our knowledge of PIE 'religion' is limited, Proto-Indo-European deities form the archetypal basis for many of our most beloved and feared gods and goddesses. Traces of these prehistoric divinities remain remarkably present in our own daily lives.

The most well-attested deity in Proto-Indo-European religion is the Sky Father *Dyeus Pater*, the earliest 'All-Father' deity in Eurasian mythology. With the spread of Indo-European culture and religion, he transformed into the Vedic deity *Dyauspitar*, the Greek *Zeus Pater*, and the Roman *Jupiter*, as well as the Germanic war gods *Týr* (in Old Norse) and *Tīw* (in Old English). All of the Proto-Indo-European deities gradually picked up their own distinctive characteristics through interactions with indigenous pantheons, with Zeus, for example, becoming an anthropomorphic 'king of the gods' in the Greek world and laying the foundations for a theological framework in which the entire world would be governed by a supreme male deity.

Proto-Indo-European societies were themselves fiercely patriarchal, but at least five major PIE goddesses can be attested through linguistic evidence, including a benevolent Earth goddess, who births and nurtures all living beings, and a goddess of the dawn.

Creation, Kinship, and Mortality

The Proto-Indo-European creation myth demonstrates the kinship between beings. In the beginning, there was voidness – neither Earth nor the heavens had yet come into being, nor had mountains, stars, the moon, or the seas. From this state, the two brothers Manu and Yemo emerged, usually accompanied by a primeval cow. Manu, the first man, kills his brother in the very first act of sacrifice, offering Yemo's body to the Sky Father in the primordial rite of worship. Thus Manu, establishing himself as the first priest, dismembers Yemo's body and uses it to fashion the Earth, sun, moon, stars, the elements, and the three social classes of humans. Priests were formed from his head, warriors from his arms, and commoners from his genitals and legs.[10]

In some versions of the story, like the Old Norse account in Snorri Sturluson's *Prose Edda*, the sacrifice of Yemo (here named Ymir) serves as the progenitive spark for *all* life on Earth:

> *Of Ymir's flesh | the earth was fashioned.*
> *And of his sweat the sea;*
> *Crags of his bones, | trees of his hair.*
> *And of his skull the sky.*[11]

It's worth noting that, in the Norse tradition, humans are considered to be the offspring of *plants*. When Odin and his brothers stumble across two trees growing along the seastrand, they use them to fashion the first couple, Askr and Embla.[12]

Kinship established by a mutual progenitor is a common feature of many global cosmogenesis myths, including in indigenous cultures far removed from the Indo-European cultural sphere. An important Māori story, for example, recorded by John White in the late 19th century, tells of the time when Tanē, god of the forests and birds, first planted trees on Aoteamua (New Zealand). He began by planting them with their 'feet' in the ground, as 'trees at first were like men,' but he found this displeasing and flipped them upside down, planting their 'heads' in the earth instead. With their tangled hair becoming their 'roots,' Tanē pronounces this 'good' and goes on his way.[13]

It's likely that stories such as these have been told for hundreds of thousands of years, enabling us to mythologically and psychologically *process* the 'mutuality of being' between human and non-human entities.[14] This is all, of course, a poetic interpretation of what we now *know* to be scientifically true: All carbon-based life on Earth *is* genetically related – and even trees and humans have far more in common than we might realize.

The Vedic World

As Indo-European traditions spread across Eurasia, they intersected with an array of indigenous worldviews, languages, and spirit pantheons. One of the earliest offshoots of the Indo-European 'family tree' were the Indo-Aryan speakers, who moved into the Indian subcontinent around 1900BCE following the collapse of the Indus Valley civilization. Indigenous peoples had populated the Indian subcontinent for tens of thousands of years, and interactions between these cultural groups and the Indo-European 'settlers' led to the eventual development of the *Vedic* tradition.[15]

The *Rig Veda* comprises the oldest layer of this ancestral tradition. It is one of the first bodies of myth ever composed in an Indo-European language. Though it was likely composed in the Punjab region of north-western India and Pakistan, its influence quickly spread across the northern Indian subcontinent and facilitated the establishment of a new 'Indic' worldview.[16]

When it comes to perceptions of non-human organisms, both the *Rig Veda* and the later *Mahābhārata* retain distinctly animistic components. In these works, we can find not only a continuum of 'spirit beliefs,' but also a lucid reckoning with the vitality of plants and animals. One account mentions a debate between the sages Bharadvāja and Bhrigu on plant sentience, where Bhrigu draws upon Vedic medical theory to demonstrate that plants engage in physiological and sensory processes based on the five elements, thus establishing them as 'sentient beings.' He states:

> ... *As one can suck up water through a bent lotus-stalk, trees also, with the aid of the [internal] wind, drink through their roots. They are susceptible [to] pleasure and pain, and grow when cut or lopped off. From these circumstances I see that trees have life. They are not inanimate... In the bodies of all mobile things [i.e. sentient beings], the five elements occur.*[17]

But despite the animistic spirit of early Vedic religion, theological and 'spiritual' evolutions in Brahmanical tradition ultimately led to perceptions of a universe mired in divine conflict, and concern for plants and the divinities of the natural world were gradually replaced with loftier and more supernatural pursuits. A distinctly malevolent force was beginning to take shape, which would ultimately impact our perceptions of *evil* on a global scale.

Picking Sides

The *Rig Veda* spoke of two major kinds of gods – the *devas* (literally 'divinities') and the *asuras* ('lords') – though these categorizations were rather loosely defined. Both were often connected with the natural world, and were regarded as morally ambiguous entities. But in later works like the *Mahābhārata*, they were pitted against each other in a great cosmic rivalry, with the *devas* being reframed as praiseworthy divine protagonists and the *asuras* as aggressive and chaotic *demons*. Thus, these important and often *ecologically embodied* deities became shunned as rebellious *anti-gods*. *Some* were legendarily 'converted' to *devas*, and ultimately new 'Hindu' gods like Shiva and Vishnu supplanted the Vedic divinities altogether,[18] but even to this day, *Sanātana Dharma* practitioners generally regard the *asuras* as a malevolent force.

Many scholars suggest that these later layers of divine conflict in the Vedic tradition were used to *allegorically* represent conflicts between India and the Persian Achaemenid Empire, which conquered Gandhara in the 6th century BCE.[19] While the *devas* became widely regarded as the gods of the Indian peoples, the *asuras* were associated with the Persians, and thus also with malice and conflict. This theory is quite compelling, particularly given the developments occurring in Persia at the time. The Persian cultural and linguistic sphere was also informed by Indo-Aryan traditions, but Zarathustra, the founder of Zoroastrianism, flipped the Indian script on its head, attesting that the *daevas* (i.e. *devas*) were in fact the evil demons to be rejected, while the *ahuras* (i.e. *asuras*) were the venerable divine protagonists.

For Zoroastrians, Ahura Mazda was established as the supreme God and 'Lord of Wisdom' and was eternally locked in cosmic conflict with Angra Mainyu, the 'Destructive Spirit,' and his cabal of evil *daevas*. The entire Zoroastrian universe was enmeshed in a struggle between 'light' and 'darkness' and this set an important precedent for later and more influential traditions.

Despite the theistic overlap, Brahmanical and Zoroastrian religious movements often came to very different conclusions about the true nature of the cosmos, including what happens after death. Early Indic traditions established a concept of *reincarnation* within the natural world, but Zarathustra opted for a starkly dualistic and transcendent afterlife for his Persian followers, constructing one of the earliest examples of a *paradise–hell* dichotomy.[20] After death, humans who followed Ahura Mazda's light would end up in paradise, while those who fell into Angra Mainyu's darkness would be transported to hell, where they would suffer until the 'Renovation of the Cosmos,' when all human souls would reunite with Ahura Mazda.

The New God from the Levant

Despite the proliferation of Indo-European mythology and cosmology, their impact on modern religious traditions was largely eclipsed by teachings from an entirely different cultural tradition.

Judaism is well known as the religion of the Israelites, whose cultural, linguistic, and spiritual traditions arose from ancient Canaanite societies stretching back to the Bronze and Iron Ages. The tribal god of the early Israelites, originally identified as El (as in *El*ohim and Isra*el*), was once one of *many* gods worshipped by ancient Canaanite peoples.[21] It wasn't until the 6th century BCE, when El was combined with Yahweh (previously a more minor god of weather and war) to form a new supreme God, that Judaism began to identify itself as a strict monotheism.[22]

The first five books of the Hebrew Bible, known as the Pentateuch, were a relatively late addition to the Jewish corpus, 'discovered' during the renovation of the Temple in 622BCE alongside the so-called 'Deuteronomic Histories.'[23] These texts emerged at a pivotal moment in ancient Jewish history. Until this time, Jewish communities maintained two small kingdoms along the eastern coast of the Mediterranean – Judah in the south, with Jerusalem as its capital, and Israel in the north, with Samaria as its capital. In the 8th century BCE, the Assyrian king Sargon II relocated 27,000 Israelites to Assyria, replacing them with settlers from his own kingdom. Many Israelites fled to Judah, causing Jerusalem's population to increase *fifteenfold* by the 7th century BCE.

When trouble at home forced the Assyrians to pull out of the region altogether, a political vacuum emerged that the now-mighty Jerusalem sought to fill. It is at precisely this moment that an entirely new layer of sacred scriptures emerged (and others were quietly removed), establishing both a single unifying God and a compelling argument for the reunification of the legendary 'Kingdom of David.'[24]

There were many mythic inspirations for the Hebrew Bible, including ancient Canaanite traditions and Babylonian tales like the Epic of Gilgamesh. Many of the 'historical' events mentioned *after* 622BCE can be corroborated in Egyptian and Assyrian sources, creating a sense of perceptual continuity between the mythic past and recorded 'history.' But the stories of Moses, the Exodus from Egypt, Joshua's conquest of Canaan, and the reign of King David are wholly unattested in the archeological and historical record.[25]

The young king of Judah at the time of the scriptures' discovery, Josiah (*c*.648–609BCE), is reported in biblical accounts to have imposed religious reforms that eradicated the worship of multiple gods, presumably to quell 'apostasy' and 'heresy' in the Jewish world. Even though polytheism had never been perceived as 'heretical' in ancient Israelite religion, Josiah professed that the Jewish people had always been *expected* to be monotheistic, and had merely allowed their covenants with their god to degrade. This argument was bolstered by the revelation of new 'histories' which demonstrated that all the past misfortunes of the Israelites were divine retribution for their worship of 'false idols.'[26]

While his influence on Israelite religion was profound, Josiah's ambition to unite the ancient kingdom ultimately never came to fruition in his lifetime. In 609BCE, he was assassinated by the Egyptian pharaoh Necho II, and within two decades Jerusalem fell to the Babylonian king Nebuchadnezzar, whose armies reduced the city to rubble. This marked the beginning of the 'Babylonian Exile' period that led to the formal codification of 'Judaism' as it's known today.

To explain why such a pious and clear-visioned king as Josiah would be forsaken by his God while clearly trying to do his bidding, his death became seen as an act of divine mercy. It's said that the God of the Israelites granted him a merciful escape from 'all the evil which I [Yahweh] will bring upon this place,' classifying the Babylonian Exile itself as yet another divine punishment for past generations of apostasy.[27]

The shift from 'gods' to 'God' in Judaism was momentous for many reasons, not least because the deities that were now seen as 'false' and 'diabolical' had once served as important intermediaries between the Canaanite peoples and their environment. Monotheism served to consolidate and centralize religion, and was an important developmental step in actualizing its social potential.

After Cyrus the Great, founder of the first Persian Empire, conquered the Neo-Babylonians in the Battle of Opis in 539BCE, Babylonian Jews were finally permitted to return to Jerusalem, ultimately leading to the construction of the Second Temple in the reign of Darius the Great. Darius himself was a devotee of Ahura Mazda, and Jerusalem remained under the Persian Empire's control until Alexander the Great conquered the region in 332BCE, beginning the process of Hellenization that ultimately led to the emergence of a new world religion.

Formation of a New World Religion

In ancient Greece, the Indo-European pantheon converged with pre-Indo-European Mycenean deities and became the Greco-Roman pantheon that we know today. By the beginning of the Common Era, most people in the Greco-Roman world adhered to a state-sanctioned form of polytheism based on the classical pantheon. But the Mediterranean remained a largely syncretic and pluralistic environment, with many people privately engaging in an array of novel religious practices.

Certainly the most impactful cultural subgroup to emerge were the Hellenic Jews, who read, spoke, and conducted rituals in Greek using a translation of the Hebrew Bible known as the *Septuagint*.[28] They had a strong presence in the Roman cities of Antioch, Alexandria, and Rome, but as time went on a schism began to emerge between the traditionalists in Jerusalem and the emerging cosmopolitan communities to the west. Further complications emerged with the rise of Jewish Apocalypticism, an array of movements centered around the idea that the end of the world was imminently approaching.

In the 1st century CE, stories of a charismatic rabbi from Nazareth began to circulate around the Hellenic Jewish world. This rabbi was, like many of his contemporaries, a propagator of Jewish Apocalypticism and, while he failed to accomplish the core messianic aim of restoring the ancient kingdom of David, his disciples expected that his return was imminent and that he would usher in the Apocalypse and definitively establish the Kingdom of God on Earth. Of course, the Apocalypse never came, forcing early Christians to shift the focus away from an imminent earthly kingdom to the promise of an *eternal* kingdom in the afterlife.

One of those early Christians, the apostle Paul, famously never met Jesus in the flesh, but it's due to his efforts that the Christian textual tradition came into being. He composed many of the most important works of the New Testament, including the Letters of Paul, which were composed *before* any of the canonical gospels.[29] A cosmopolitan reformer, he sought to bring a *new* kind of Judaism to the gentile public. A figurehead like Jesus was a perfect conduit for such an endeavor.

By severing the 'Jesus movement' from its ethnoreligious roots and allowing it to assimilate other worldviews, Paul enabled his burgeoning new faith to become one of the first true 'world religions.'[30] The iconographic and linguistic *Hellenization* of Jesus and his teachings, along with the cultural input of figures like Paul, further facilitated a firm connection between Christianity and Greek Mediterranean culture, establishing a persisting idea of Jesus as an exemplar of 'western' values.

While Christianity was only one of many religious movements to arise in the Greco-Roman world, it clearly had the most enduring impact. This was due in large part to its installation as the state religion of Rome, which, although ultimately detrimental to the empire itself, did manage to give Christianity substantial political clout. In the 4th century, bishops were paid six times as much as doctors and the Church was wholly exempt from taxation.[31] Engagement with religion was a lucrative path to the upper echelons of society, where pious Church Fathers could indulge in the luxuries of worldly life while awaiting their eternal paradise in the hereafter.

The scale of cultural erasure in the early Christian era is often sorely underacknowledged. If you've ever wondered why there are so many headless statues from the classical era, this is precisely the cause. Statues were not only decapitated, but drawn, quartered, and defiled by pious religious zealots. Altars were smashed, manuscripts erased, and beloved icons dismembered before the eyes of their devotees.[32] To unite everyone under one religious authority, it was necessary to vilify the old gods and destroy any trace of their worship.

Of course, not all of those who followed the teachings attributed to Jesus were part of Paul's Christian movement. Jewish sects like the Nazarenes and Ebionites observed the laws of the Torah while following an Aramaic translation of the Greek Gospel of Matthew, and their traditions endured quite independently for the first four centuries of the Common Era. While groups such as these were ultimately overshadowed by the Christ movement, remnants of their legacy can still be found in the later Islamic tradition.[33]

Genesis and Dominion

In all three religions of the Abrahamic world, Judaism, Christianity, and Islam, mankind is clearly established as the existential center of the universe. The first chapter of Genesis states:

> And God said to them, 'Be fruitful and multiply and fill the earth and subdue it, and have dominion over the fish of the sea and over the birds of the heavens and over every living thing that moves on the earth.' And God said, 'Behold, I have given you every plant yielding seed that is on the face of all the earth, and every tree with seed in its fruit. You shall have them for food. And to every beast of the earth and to every bird of the heavens and to everything that creeps on the earth, everything that has the breath of life, I have given every green plant for food.' And it was so.[34]

While some translations seek to soften this passage, 'subdue' and 'dominion' are quite faithful to the original Hebrew. Some eco-critics have pointed to this single line of text as being one of the most consequential statements ever written regarding humanity's relationship with nature. In 'The Historical Roots of our Ecological Crisis,' Lynn White Jr writes:

> Especially in its Western form, Christianity is the most anthropocentric religion the world has seen... Man shares, in great measure, God's transcendence of nature. Christianity, in absolute contrast to ancient paganism and Asia's religions (except, perhaps, Zoroastrianism), not only established a dualism of man and nature but also insisted that it is God's will that man exploit nature for his proper ends.[35]

This theological construct was quite easily absorbed into a Greco-Roman world that was already primed by the anthropocentricity of Aristotelianism, leading to a confluence of 'rational' and 'spiritual' justifications for ecological disregard.

Divinity, Evil, and Nature

Our conceptions of the *divine* can have a deep impact on the way we relate to nature. Ideas of divinity can be used to foster a sense of reverence for the ground beneath our feet or push us toward ecological dissociation. This is a common feature of many forms of modern religion, where ascetic practices like celibacy are used as a reminder that the physical world is base and contemptible. The idea that divinity or enlightenment can only be reached through complete *dissociation* from materiality is also bound to foster a relative disdain for the dynamics of nature.

In Mediterranean societies, the spiritual rejection of the 'baseness' of the Earth became particularly prominent with the rise of Gnosticism. Gnosticism was a mystical Judeo-Christian movement in the early centuries CE, which experienced a brief revival in the medieval era and another in recent centuries. While rooted in the Christian tradition, it can be best described as a spiritual path dedicated to the pursuit of *gnosis*, or 'mystical knowledge,' cultivated through direct *experience* of God rather than orthodox religious practices. The Gnostics believed in a starkly dualistic universe, presided over *not* by the supreme transcendent God (who was 'hidden' and accessible only through the divine intellect), but by an evil intermediary, a *demiurge*, whose malevolence was expressed in his diabolical creation of an 'evil' material universe.

Notions of a demiurge stretch back to ancient Greece, where Plato posited a *benevolent* demiurge as the intermediary creator of the material world. But the Gnostics challenged the benevolence of the world's 'creator,' and in the Cathar revival of the 12th to 14th centuries, the demiurge became identified with the Devil himself. Gnosticism largely fizzled out in Europe during the late Middle Ages, but associations between *nature* and *evil* retained their appeal. Over time, our gods became more and more *transcendent*, while demons grew increasingly *earthly*. While divinity was projected into the distant cosmos, our own world became seen as the domain of the Devil.

It should be noted that notions like hell and the Devil were all late additions to the Abrahamic faiths, likely arising out of interactions between Jewish Apocalypticism and Zoroastrianism.[36] Like any mythic character, the Devil evolved over time. While the title *śāṭān* is occasionally used in the Hebrew Bible, there it simply meant 'accuser' or 'adversary,' usually referring to distinctly *human* enemies. In one case, it is used to refer to an angelic prosecutor in service to Yahweh, but this divine administrator bears little resemblance to the omni-malevolent Christian Devil, who was never a feature of mainstream Jewish religion[37] Our own word 'devil' is derived from Greek *diabolos* ('accuser' or 'slanderer'), a literal translation of Hebrew *śāṭān*.

And what about demons? Where do these minions of darkness come from? In the Hebrew Bible, what we term 'demons' can be divided into two major classes: the *se'irim* and the *shedim*, both referring to the 'false gods' once worshipped by the ancient Israelites. In the *Septuagint*, *shedim* was translated as *daimonia*, an old Greek name for powerful spirits. Over time, this category of *daimonia* came to include many gods of the classical 'pagan' world, who were stripped of any venerable qualities and transformed into archetypes of evil. These deities, who once represented the spirited core of the natural world, were now not only rejected as fictions but condemned as *demons*. As Augustine once pronounced, 'All the pagans were under the power of demons. Temples were built to demons, altars were set up to demons, priests ordained for the service of demons, sacrifices offered to demons, and ecstatic ravers were brought in as prophets for demons.'[38]

While the Devil himself was a rather hapless figure in much of the early Middle Ages, his reputation began to change dramatically in the wake of the Black Death. People started to wonder whether plague might itself be the handiwork of a far more active and dangerous Devil – no longer merely an amorphous 'deceiver' but a formidable force of nature.

But perhaps the most impactful revolution in western perceptions of devils and demons occurred in 1486, with the mass publication of a phenomenally important book. Not the Bible, which was printed by Gutenberg 30 years prior, but the *Malleus Maleficarum – The Hammer of Witches*. This famous witch-hunting guide was written by a German inquisitor, Heinrich Kramer, after returning to Germany following a tumultuous mission in Innsbruck.

When he arrived in Austria in 1485, he encountered a woman named Helena Scheuberin. She was no fan of inquisitors, and made her distaste for Kramer's presence known as soon as he arrived in her mountain town. When he passed her in the street, she reportedly spat on him, shouting, 'Fie on you, you bad monk, may the falling evil take you.'[39] She became a thorn in his side, publicly boycotting his sermons and encouraging others to do the same, so Kramer decided to take her to court on allegations of witchcraft. He claimed she had used diabolical magic to kill a local knight, substantiating his theory with allegations of her sexual impropriety. But the town's investigating commission failed to see how such evidence was indicative of witchcraft, and moved to acquit Scheuberin.[40]

But Kramer refused to relent. He lingered in Innsbruck for months, harassing local townspeople for evidence against Scheuberin, ultimately forcing the local bishop to expel him and send him back to Cologne.[41] Humiliated and indignant, it was there that he set pen to paper and composed the *Malleus Maleficarum*. The book was both sensational and controversial, initially condemned by many of Cologne's top theologians as a departure from established Catholic doctrine.[42] The Devil was known to be a liar and deceiver, not a bestower of supernatural powers. But with time, works like the *Malleus* effectively convinced European Christians that malevolent magic was a serious threat – and predominantly the domain of women.

Unlike previous treatises on witchcraft, the *Malleus* makes a concerted effort to establish direct links between witchcraft and female sexuality, claiming that demons used sex to lure immoral women into witchcraft. It was a slippery slope from promiscuity to demonic sex rituals and infant sacrifices for the preparation of magical potions.[43] Even a miscarriage or abortion could be grounds for a death trial. Rather than simply being deceived by Satan, women were perceived as colluding with him in a diabolical conspiracy against God.

Between the 16th and 17th centuries, approximately 40,000–60,000 people were executed during the witch-trials. Of these, around 80 per cent were women.[44] Naturally, we can safely assume that none of these women were *actually* in magical league with the Devil, nor did most have any direct involvement in activities we would now associate (often romantically) with 'witchcraft.' Simply being a woman was enough to justify suspicions of *maleficium*. But there was certainly also a continent-wide crackdown on 'pagan' spirituality at this time, as well as a marked increase in anti-Semitism.[45] Kramer also formalized notions of animal 'familiars,' leading to a mass culling of domestic cats and other animals out of the fear that they were the non-human associates of witches.

Overall, the European witch-trials signalled a monumental shift in Christian perceptions of evil. Rather than being the 'fantasies of overwrought human minds,'[46] the threat of witchcraft made the workings of the Devil a tangible and imminent danger – directly tied to the Earth, natural magic, and femininity.

This new take on 'evil' was enthusiastically adopted in the Protestant Reformation by Martin Luther,[47] who further established the Devil as an ever-present threat, lurking in the shadows of our minds, our societies, and the natural world itself.

The Dogma of Disregard

While the ecological pitfalls of viewing nature as 'evil' should be clear, we must be wary of thinking that *deification* is any better. One of the most dangerous beliefs we can hold is that our behaviors *can't* meaningfully impact the planet, either because it's *divine* or because it's under the absolute control of an omnipotent God. Trees, animals, and tropical rainforests are not immortal deities – they are sentient beings who suffer and die, and it would be a mistake to think that our deeds cannot harm them.

Deification can be a particularly nasty form of 'othering,' which we can easily see in our treatment of 'royals' and 'celebrities' in modern society. We can also see the pitfalls of sanctity in fundamentalist religious traditions that reduce female-bodied individuals to 'sacred vessels for procreation,' leading to the subordination and micro-management of their affairs by the 'holy men.' The reduction of living beings to any form of sacred archetype does very little to establish a baseline of empathy or respect.

When we deify other beings, we rob them of both their agency and their *vulnerability*. Once, when I was living in Kathmandu, I found myself sitting on the banks of the Bagmati river at Pashupatinath, just above the platforms where cadavers were washed and prepared for cremation. I noticed a devotee walking down to the water with a packet of incense. I watched as she unwrapped the bundle of sticks and tossed the plastic wrapper into the sacred river, then lit the incense and recited her prayers without a second thought. This apparent dissonance is unfortunately all too common, even in societies where religion and nature are closely intertwined. It's clear that the impacts of *deification*, *sanctification* and *personification* are quite distinct, largely because *personhood* alone provides space for vulnerability.

But over time, some of the world's most powerful religions came to imagine a direct correspondence between nature and evil, with the 'natural' associations of femininity leading to the increasing oppression of women. This was in line, in many ways, with the primary Platonic division between the superior *masculine* realm of reason and the inferior *feminine* realm of physical nature. Even in 'wisdom traditions' like Gnosticism, the living world was seen to be no more than a diabolical illusion, obscuring the purity of the transcendent God. Most consequentially, these systems helped to sanctify the notion that humans were the sole biological organisms with intrinsic existential value and to sunder 'humanity' from the perennial *other* of 'nature.'

The sanctification of anthropocentrism was deeply influential, both because we tend to be quite emotionally attached to religion and because the formation of 'world religions' allowed such ideas to spread far beyond their original cultural context. It's one thing to say that 'some Greek philosopher' thought that humans were superior to animals, but quite another to say that the omniscient creator of the heavens and the Earth declared it himself.

Whether we like it or not, religious myths such as these still dictate a great deal of what we think of as 'moral' behavior. We turn to religion to shed light on the meaning of life and our place in the universe. But religious ideas are always unavoidably a product of their own time and place. No religious authority, be it textual or incarnate, is non-negotiable or infallible.

This isn't to deny the evolutionary power of religion. Through it, we can form powerful social bonds that radically increase our odds of collective survival. But, in a diverse and multicultural world, an exclusivist approach will always give rise to division. If religion is going to endure as a force for good in the world, it will need to become far more comfortable with pluralism and ultimately expand beyond anthropocentric delusions to consider all *beings*.

Part III
PROGNOSIS

6

PROVOCATION

'Spirit Illness' and Ecological Health

We are at a profound turning-point in the progression of our ecological disease, and our prognosis depends entirely on the decisions we make surrounding our treatment. There *is* still hope of recovery, at least *for some*, but things remain highly precarious. We must remember that humanity is not an undifferentiated collective – not all humans bear the same responsibility for climate change, nor will all of us experience the same consequences. Those with greater means and privilege (and often also greater culpability) will naturally have a greater capacity to protect themselves from things like drought, food shortages, extreme heat, or rising sea levels. We will *all* ultimately face the consequences of anthropogenic climate change, but they won't be proportionately or evenly distributed.

Pandemics, of course, are somewhat of an exception to this rule. They don't particularly care about wealth, class, or human hierarchies. While there are always social factors at play in pandemics, diseases like Covid-19 should remind us that disease can be a powerful equalizing force – and that some of the most harrowing consequences of the Anthropocene will have massive global effects. It's unfortunately rare that we speak of pandemics as having much to do with climate change, but as we'll see, the two are quite intimately related.

The Covid-19 pandemic has been an era-defining experience in many ways. While some important advances came out of this tragedy, it's estimated at the time of writing that over 15 million lives have been lost to Covid-19 so far.[1] This number pales in comparison to projections of where we were heading *without* vaccines, which are estimated to have averted nearly 20 million Covid deaths in 2021 alone,[2] but grappling with our collective trauma and loss has itself become a source of trauma. Social scientists have reported distinct shifts in the ways that we culturally deal with grief in the wake of the Covid pandemic. Public expressions of mourning (for instance, on social media) are routinely met with criticism, ridicule, and conspiracy.[3] While one might expect that a global tragedy would bring us all together and make us more empathetic to the needs of others, in effect it has made many of us *more* individualistic, more callous, and more vitriolic than ever before.

There are many lessons that we can glean from the past few years. We have hopefully come to realize that data alone are not always sufficient to compel a shift in human behavior. The ever-widening rift between the hard sciences, social sciences, humanities, and arts has left us with a crisis of disintegrated knowledge. At the end of the day, data need to be both understood and psychologically integrated to motivate a change in behavior. The path to effecting such a change necessarily involves research, data analysis, an understanding of social and historical context, and a creative capacity to make knowledge accessible to a diverse audience. But traditionally, interdisciplinary approaches of this sort are few and far between. It's not common that you get scientists, historians, sociologists, philosophers, storytellers, and politicians in a room together to discuss solutions to world issues, but this is indeed the kind of dialogue that is necessary. One of the most important lessons to be learned from Covid is that we need far more active collaboration between the sciences and humanities to guide us toward a truly holistic understanding of our world.

Unfortunately, the emergence of this dire symptom was all too predictable and is likely to represent a new normal in our years to come. Scientists have been warning us for decades that climate change will inevitably lead to an uptick in novel infectious diseases. Part of this is a consequence of shifting ecologies, melting permafrost, and an expansion of climatic conditions in which pathogens thrive. But it is also a direct consequence of our transgression against the boundaries of the natural world, and our troubled relationships with non-human beings. As we venture deeper into the wilderness in pursuit of profit, we inevitably end up exposing ourselves to a plethora of unfamiliar beings and foreign microbiomes. In places where 'exotic' animals are captured and slaughtered for consumer goods, these exposures can lead to devastating and rapid consequences. But even in our own backyards, industrial animal agriculture operations are a breeding ground for far more than livestock, and outbreaks of swine and bird flu can quickly snowball into a public health disaster. If we fail to take the threat of pandemic infections seriously and address the clear links between infectious disease and ecology, then Covid-19 is unfortunately bound to be just the tip of the pathogenic iceberg.

Pandemonium

Pandemics are obviously not a new thing, but they're also not *that* old, geologically speaking. Most of our familiar infectious pathogens only adapted to humans in the past 12,000 years with the proliferation of animal agriculture, though their impact in that time has been impressively catastrophic.[4] While most of us view the Agricultural Revolution as a great leap forward in human civilization, our exploitative and unwary approach to animal-rearing created prime circumstances for disease to develop and spread to us, and, due to the more settled and concentrated nature of agrarian societies, these infectious pathogens could continuously spread and evolve unchecked for thousands of years.

In ancient times, medical systems around the world took an array of different approaches to dealing with infectious diseases. In many ancient traditions, pandemics were understood to be caused by *unseen beings* like demons, nymphs, spirits, and other supernatural forces. But in the Greco-Roman world, this came to be thought of as superstitious nonsense. Medical theorists like Galen postulated that 'bad air' (*miasma*) was the *true* culprit behind pandemics. This theory gave us the distinctive beak-like masks worn by plague doctors, which were filled with fragrant herbs to mask the infectious stink.

While Galenic medicine gained significant momentum across Eurasia, doctors in other parts of the world often came to rather different conclusions regarding the nature of infectious disease. Tibetan Medicine provides a particularly poignant example. When Indian, Chinese, and Greco-Arabic medicine began to trickle into Tibet around the seventh to ninth centuries CE, early Tibetan systematizers sought to blend their indigenous medical knowledge with these seemingly disparate traditions to form a unified 'Science of Healing' (*Sowa Rigpa*, Wyl. *gSo ba Rig pa*). This was ultimately formalized in the 12th century with the composition of the *Four Tantras* (*Gyü-Zhi*, Wyl. *rGyud bZhi*) by Yuthok Yönten Gönpo. This highly venerated corpus remains the theoretical and pedagogical basis for Tibetan Medicine to this day.

Systems like *Sowa Rigpa*, Ayurveda, and Chinese medicine are often referred to as 'traditional' medical systems. When we hear the word 'tradition,' we tend to think of 'the old way of doing things.' As such, traditional medicine is often characterized as a kind of *resistance* against modernity, or an *alternative* to progress. But 'tradition' really just means 'lineage' in this context. Traditions such as these were cultivated carefully over time, with input from many different sources and many generations of practitioners. Ancient medicine was nearly always a multi-cultural affair; illness was too serious a matter to be dealt with in an insular fashion. It was far more pragmatic to observe what other people were doing, give it a try, and, if the results were favorable, find a way to convince other people to try it too. On this basis, many systems of traditional medicine made a concerted effort to share 'notes' and substances with others to find the most effective treatments for their patients.[5]

In the Mediterranean world, the movement away from a spirited Earth underlaid many of the medical advances of Hippocratic and Galenic medicine. Galenism certainly acknowledged *environmental* factors in health and disease, but these were strictly devoid of vital agency. Of course, we now know that non-human organisms actually have a *massive* impact on human health, but for a long time in Europe this idea was considered ridiculous.

While Galenic medicine spread widely throughout Eurasia, Asian medical traditions generally came to very different conclusions regarding the role of the spirited environment in disease etiology. Our *relationships* with the beings of the natural world, both seen and unseen, were understood to be integral components of our individual and social health.

Meeting the Science of Healing

While studying at Naropa University in Boulder, Colorado, in 2009, I became quite ill. I was vomiting multiple times a day for months on end, and with no helpful answers coming from the network GP available through my student insurance plan, I decided to look online for other options. I had long been interested in Tibetan Medicine, having read a primer on the subject when I was a teenager. I even managed to order some Tibetan pills online while I was in high school in a desperate attempt to control my acne. But I had never lived in an area where finding a Tibetan doctor was even a remote possibility until I moved to Boulder.

A friend recommended that I book an appointment with Dr Nashalla Gwyn-Nyinda, a western *menpa* (Wyl. *sman pa*, Tibetan doctor) who had trained both in the Himalayas and at the Shang Shung Institute in Massachusetts, the latter being one of the few institutions to offer Tibetan Medical certifications outside Asia. I was thrilled to *finally* meet a Tibetan doctor, and looked on in awe as she carefully read my pulse and examined my urine sample. After talking a bit about my symptoms, diet, and lifestyle, she presented her diagnosis, set me up with a tailored nutritional and behavioral plan, and sent me on my way with a few bags of little brown pills. Within a week, the symptoms that had plagued me for months had completely disappeared, and by the time I returned for a check-up, I felt better than I had in years. I was astonished by the efficacy of these methods and therapies, and desperately wanted to learn more.

When an announcement popped up just a few weeks later that the Shang Shung Institute was accepting applications for a new cohort of students, I applied without hesitation. Fortuitously being accepted, I left my religious studies program behind and began a very long love affair with Tibetan Medicine. The *Four Tantras* comprised the core of our curriculum, with in-depth explanations and commentary provided by our professor, Dr Phuntsog Wangmo, along with practical training in diagnostics, external therapies, herbal medicine, and an array of other topics. While my years at Tibetan Medicine school were among the most challenging in my life, they were also profoundly enriching, and by the time I graduated I found that my experience of the world had dramatically shifted.

Outside of my program, I also began studying with Dr Nida Chenagtsang, a yogi-physician from north-eastern Tibet. As a lineage-holder of the *Yuthok Nyingthig*, a Tibetan Buddhist practice cycle associated with Tibetan Medicine, Dr Nida has been instrumental in bringing both the 'secular' and 'spiritual' aspects of Tibetan Medicine to a global audience. In 2017, he sent me to Nepal to complete a five-month medical internship, and in the years that followed I began teaching on *Sowa Rigpa* and the *Yuthok Nyingthig* under Dr Nida's aegis.

The Causes of Illness

Buddhist philosophy asserts that all experiences of suffering, including illness, are a primary result of the mental afflictions of *ignorance, attachment,* and *aversion,* which themselves arise due to our *unawareness* of the true nature of reality. All illness is therefore seen as the inevitable outcome of the fundamental rift between our worldview and the way things truly are – a common experience, according to Buddhism, for all sentient beings.

On a more pragmatic level, however, Tibetan Medicine specifically acknowledges four proximal *conditions* that propel us into an imbalanced state. These are *diet, lifestyle, time/season,* and *provocation.*

Diet and *behavior* are viewed as both *causes of* and *treatments for* disease. As two of the primary ways in which we engage with the world around us, they can foster a state of dynamic balance *or* make us deeply unwell. *Time* and *season* can also cause cyclical and environmental fluctuations in our health, reflecting both our natural 24-hour biorhythm as well as the impact of seasonal changes on our physiology and epidemiology. While many seasonally oriented imbalances are *also* linked to our diet and lifestyle, their *chronological* dynamics remind us that we are all part of an interdependent world in a constant state of flux.

But for the discussion at hand, it's the final factor that's most important – 'provocation,' or *dön* (Wyl. *gdon*). Rather than being caused by diet, lifestyle, or chronological cycles, *provocation*-based disorders are specifically caused by *other beings,* namely those who are conventionally *unseen.* Within this category we can find a wide range of infectious diseases caused by microscopic pathogens, as well as numerous other physical and mental afflictions. In all cases, these states of imbalance are either directly or indirectly caused by unseen beings who have been 'provoked' by human disturbances.

This theory is notably distinct from some of the other ostensibly 'supernatural' disease paradigms found in the ancient western world. In their assault on superstition, Hippocrates and Galen were principally reacting to the idea that disease was a punishment sent by the gods, positing instead that it was caused by physiological and environmental factors. Buddhist medical traditions of all kinds had already made this leap. In Buddhism, while our individual *karma* is understood to play a role in *all* of our experiences, thus potentially making the argument that illness is a result of 'bad karma,' this has nothing to do with 'gods' or 'punishment,' and is merely a latent background factor (as with any experience of suffering) when it comes to most Buddhist approaches to disease.

Indian Ayurveda was an important source of Buddhist approaches to 'spirit illness' in Tibet, particularly the sections on the treatment of spirit/ghost disorders found in the *Aṣṭāṅga Hṛdaya Saṃhitā*. This mighty pillar of Ayurvedic theory was composed by the doctor Vāgbhata around 600 CE and translated into Tibetan in the 11th century, after which it quickly gained traction in Tibet as an important and reliable basis for medical theory. In this text, as in numerous earlier Ayurvedic works, the management of spirit-related illnesses was presented as a normal part of a doctor's job. The text specifically lists 18 ghost/spirit illnesses, largely recognizable as forms of *possession*, and these were later adapted by Yuthok for his own overview of spirit possession in the *Four Tantras*.[6]

Ayurvedic literature has always been a goldmine of information pertaining to spirit illness, but most modern Ayurvedic academies steer clear of the topic, largely in an attempt to sanitize and adapt this ancient tradition to the sensibilities of a 'disenchanted'

Euro-American world. Tibetan Medicine, on the other hand, has not only managed to retain the integrity of its lineage (and indeed to build upon it) for over 800 years, but has also avoided falling prey to anthropocentric disenchantment. This could all change in the years to come – many of the teachers I've encountered, both in a medical and spiritual context, tend to operate under the assumption that staying quiet about spirits is necessary to be relevant to a western audience – but it's important that paradigms like these are not lost, as they offer valuable alternate perspectives on the inseparability of 'nature' and 'health.'

Provocation Disorders

In early Tibetan medical works, spirits are characterized as being a prominent underlying cause for a broad range of illnesses, but in the *Four Tantras*, Yuthok takes a somewhat more nuanced and conservative approach. In his view, the *majority* of illnesses result from unsuitable diet, behavior, and other conflicting factors, and only a small fraction have a spirit-related cause. Some of these so-called '*provocation* disorders' have distinctly psychiatric components, but even in the 12th century it was considered vitally important to distinguish between mental illnesses with a physiological or psychological basis and those brought about by external forces.

In Tibetan Medicine, contagious diseases are themselves considered to be *provocation-based* disorders, with many Tibetan doctors deeming the Covid-19 pandemic a natural consequence of our destructive relationship with nature. Pollution, deforestation, and disruption of the air, water, soil, trees, and sensitive natural places are all believed to produce illness among unseen beings like the sky-dwelling *mamos*, chthonic *nāgas*, and tree-dwelling *nyen*. When their sickness and indignation spill over into the biological realm, humans and other beings become ill too, and pass on their illness to other beings.[7]

The idea that unseen beings could act as vectors for disease was not unknown in western Eurasia, but it mostly fell out of favor during the 1,500-year reign of Galenism, with most 'respectable' European doctors remaining convinced that bad stenches were principally responsible for pandemics like the Black Death. While successful treatments *can* sometimes be founded on inaccurate theories, Galenic medicine was unfortunately also quite ineffectual when it came to *treating* such infections. Things didn't start to improve in Europe until the 18th century, with the adoption of variolation, an Asian and African technique that directly contributed to the invention of the smallpox vaccine and a subsequent explosion in 'western' therapeutic advancements.

Tibetan doctors ended up producing significantly more efficacious solutions for infectious diseases, and this didn't go entirely unnoticed. In the 1850s, a Russian military general stationed in eastern Siberia found himself in a grave predicament when his troop was devastated by a typhoid or cholera outbreak. He sought the help of a local Tibetan doctor, and was so impressed by the efficacy of his therapies that he ultimately invited him and his brother to work at a hospital in St Petersburg. Pyotr Badmayev, one of the brothers, went on to establish a Tibetan medical clinic and institute there, and was even appointed as a minister of foreign affairs for the Russian Tsar.[8] Tibetan Medicine came very close to being established as an official 'Russian' alternative to European biomedicine, but such developments were eventually abandoned in the wake of the Russian revolution.[9]

From a strictly biological point of view, it's clear that infectious diseases are indeed caused by 'unseen beings' – we simply refer to them as viruses, bacteria, etc. Our ability to 'see' these organisms under microscopes quickly led us to the conclusion that we had finally 'figured it all out.' But René Dubos, the researcher who isolated the first commercial antibiotic in 1939, strongly warned against becoming overly confident in our purported understanding of the microbiome. Even though he made his mark on medicine by developing technologies for eradicating microbes, his later research focused on the dynamic 'symbiosis of Earth and humankind' as the key to managing infectious diseases. Rather than fixating on a myopic, 'germ-eye' view of infections, he was one of many scientists to argue for a more holistic approach to virology and epidemiology, writing:

> Modern man believes that he has achieved almost complete mastery over the natural forces which molded his evolution in the past and that he can now control his own biological and cultural destiny. But this may be an illusion. Like all other living things, he is part of an immensely complex ecological system and is bound to all its components by innumerable links.[10]

As Mark Honigsbaum illustrates in The Pandemic Century, a complex and interconnected array of factors contribute to the outbreak of pandemics, including urbanization, increased global interconnectivity with low transit times, a growing demand for animal proteins and dairy products in the industrial world, deforestation and the destruction of wild places (mostly for animal agriculture), social and behavioral practices, deficient public health care, and many other factors. While microbes are certainly the *primary* etiological factor in diseases like Covid-19, our *relationships* with them are influenced by many interwoven conditions. Honigsbaum writes, 'Unless and until we take account of the ecological, immunological, and behavioral factors that govern the emergence and spread of novel pathogens, our knowledge of such microbes and their connection to disease is bound to be partial and incomplete.'[11]

In the Tibetan approach, the assumption of non-human *agency* in infectious disease processes creates an important space for biological dynamism. Rather than thinking of viruses as tiny assemblies of molecules, we're pushed to consider them as decision-making entities in our shared environment. In this light, the 'prevention' of epidemics becomes an ecological and social issue, not simply a matter of medical therapies.

But Buddhist medical traditions have not shied away from the immediate needs of sick patients, nor the importance of eliminating infectious invaders from the body. In Tibetan herbalism, many of the substances prescribed to expel 'unseen beings' and their pathogenic traces from the body are now well-established as having anti-microbial properties. It should come as no surprise that many such substances (like myrrh resin), famed throughout the world for their ability to repel 'evil spirits,' have also had long-standing traditional applications in the treatment of infectious disease. This isn't to say that *all* 'spirits' of illness are really just microscopic organisms, however, even though there is a clear and crucial correlation between the two.

Asian medical traditions have also made some perhaps surprising contributions to our 'modern western approach' to mitigating contagious disease. Vaccines, in particular, were originally derived from a process known as *variolation*, which was used in African and Asian medical traditions for thousands of years in managing smallpox.[12] In its most basic form, variolation simply involved taking some infectious tissue from a patient's smallpox pustules and inserting it subcutaneously into an uninfected patient. When successful, this controlled exposure would allow the body to build up a natural immune response so that serious illness could be avoided in the future. By the 18th century, this 'foreign' medical innovation had become quite popular in Europe, including among the aristocracy. It even made its way to the New World by 1721, 200 years after indigenous American communities were first decimated by European smallpox.[13] Later, in 1796, English physician Edward Jenner claimed to have 'invented' the smallpox vaccine after himself 'discovering' that milkmaids with previous cowpox infections were naturally immune to smallpox. This popular myth is, in fact, a rather egregious whitewashing of scientific history, designed to disguise the fact that vaccination *wasn't* a brilliant English medical discovery, but an adaptation of a *traditional* practice from far outside the 'enlightened' European scientific sphere.

Ecological Health and Sensitive Places

In Tibetan Medicine, *ecological* health is fundamentally inseparable from *physiological* health, both on an individual and social level. Actions like felling trees, digging untouched earth, mining stone, manipulating waterways, killing animals, and polluting the air are all understood to be *provocative* acts that can have dreadful consequences. If it's deemed necessary to break ground or fell trees, then it's common in Tibetan societies to engage in divination and ask the local non-human spirits for permission, and to follow any such acts with offerings of thanksgiving and even rites of apology. Ceremonies such as these, especially when performed in a group setting, help to psychologically and socially reinforce a sense of real *relationship* between human and non-human beings within an environment, as well as acting as a reminder of the consequences that might arise if restraint is not sufficiently exercised.

Such concepts are, of course, not *only* present in Tibet. In the Anglo-Saxon world, many healers maintained that a phenomenon known as *Elfshot* was a prominent cause of physiological distress, especially in cases of rheumatism, arthritis, and idiopathic pains. According to this theory, elves inflict suffering upon unwary humans by shooting them with invisible arrows, the purported remnants of which were often recovered from what we now know to be Neolithic tombs in the form of arrowheads and flintstones. These temperamental spirits were thought to influence a wide array of conditions, although, much like the spirits of Tibet, they weren't thought to be inherently evil (or inherently good), but merely *another form of life* with which we share the world.

In Tibet, Yuthok's *Four Tantras* are an essential source of information regarding identifying and treating provocation disorders, and Yuthok presents a pithy explanation of the kinds of ecological behaviors that give rise to such outbreaks, including:

> *Upturning Earth spirits by digging up turf and fields,*
> *Disturbing Water spirits by damming waterways and*
> *flooding meadows,*
> *Cutting down Tree spirits and uprooting Stone spirits,*
> *Performing wanton acts like burning impure substances,*
> *Contaminating the hearth [e.g. with toxic or foul substances],*
> *Killing beings in sensitive places,*

Rousing the entities of the sensitive places to subdue one's enemies with magical feats.[14]

'Sensitive' places are what we might think of as 'wild' – locations that are not accustomed to human activity, and are best left undisturbed. They are often important ecological locations, like water springs, caves, and forests, and spirits of all sorts are seen to inhabit them.

The term that I've translated above as 'spirit,' *nyen* (Wyl. *gnyan*), can have a range of meanings, reflecting the important overlap between notions of physical health, the environment, and the spirit world. In a medical sense, *nyen* refers to certain 'sensitive' and painful disorders linked to inflammation and infection, and the same term is used as an indigenous Tibetan classification for nature spirits who are particularly 'sensitive' or 'vulnerable.' These beings generally dwell in 'sensitive places,' which are thus known as *nyensa*. Sensitive places are often ecologically vulnerable and untamed locations, but any place that is inhabited (or *haunted*) by unseen beings can be referred to as a *nyensa*.

Most *nyen* are thought to live in remote mountain forests, especially coniferous ones, where they may take on the forms of an animal, like yak and sheep, or even a humanoid entity, though their 'true form' remains more mutable. It was commonly held that if a *nyensa* is disturbed (i.e. if a *nyen*'s tree is cut down), the resident spirit may become angry and inflict nearby populations with disease.[15]

It's interesting that nowadays *nyen* spirits are mostly relegated to specialist religious or medical contexts in Tibetan and Himalayan societies, while spirits like the *nāga* (known as *lu* in Tibetan, Wyl. *klu*) have far surpassed them in terms of folkloric popularity. This speaks quite directly to the interrelatedness of ecology and spirit paradigms. While the Tibetan Plateau is generally thought of as a treeless and desolate country, pollen and charcoal analyses have demonstrated that it was once widely forested with trees and shrubs like juniper, birch, willow, and sea buckthorn. An early wave of deforestation began around 7000BCE, when Neolithic nomads started burning and clear-cutting forests to create pastureland for their livestock.[16] But a second wave began in the Imperial Era, and has continued to the present day. The mountains around Lhasa were still at least *partially* forested when the Jokhang Temple was first constructed in the 7th century, and it really wasn't until the 15th century, with the establishment of massive monastic complexes around Lhasa, that the region was reduced to a treeless desert.[17]

Even in the 21st century, some near-centenarian residents of Jharkot, a culturally Tibetan region of Nepal, recall the existence of birch forests in the now treeless, alpine expanses of their early childhood.[18] Many such 'wild' forests endured into the modern era, only to slowly disappear with the rise of industry over the past century, causing their resident *nyen* spirits to likewise fade into relative obscurity in popular imagination. The *nāga* spirits, by contrast, are often more domestic and sociable – associated with parks, waterways, and cultivated lands rather than wild and untamed forests. As Jakub Kocurek argues, 'The loss of forests in Tibet has led to a gradual forgetting of the *gnyan*. On the other hand, the *klu* have become prevalent as tree-dwelling beings, since the surviving trees were mostly located near rivers and other water sources.'[19] While ancient texts do occasionally point to the *nāgas'* affinity with trees, their popularity as 'tree spirits' in the modern Tibetan world is largely a function of Tibet's changing climate.

While a simple 'belief in spirits' certainly didn't spare the Tibetan wilderness from human exploitation, it's notable that, save for the early wave of climate transformation in the Neolithic era, the deforestation of the plateau was principally carried out by foreign institutions seeking to convert and 'tame' Tibetan communities, both human and non-human. This offers an important shift in perspective when looking at stories of great masters 'subjugating' the hostile spirits of the Tibetan landscape to establish temples and monasteries. In many cases, this process of subjugation was accompanied by a very real process of ecological exploitation and degradation.

But even in classical Tibetan Buddhist sources, such behaviors were well understood to be risky. Spirits like the *nyen* and *lu* are still thought to cause physiological and psychological afflictions when their abodes are disturbed. *Nyen*, in particular, are commonly associated with cancer, thus creating a perceived etiological link between environmental desecration and this and other illnesses. Based on this rationale, many modern Tibetan doctors perceive a correlation between modern mass deforestation and rising cancer diagnoses worldwide.

While it's important to avoid destructive behaviors *anywhere*, sensitivity to the ecological and spiritual context of a place can go a long way in mitigating serious harm. In our global ecosystem, it's reasonable to think of places like the Amazon rainforest, the polar ice-caps, the Himalayan glaciers, old growth forests, coral reefs, etc., as important and highly sensitive *nyensa*, and thus localities that must be treated with the utmost respect. If our behaviors, even from afar, threaten to harm these spirited locations, then we run the risk of provoking serious harm for both human and non-human beings. With the exception of indigenously led engagements, it's often better to leave such places alone. But even when dealing with beings in our own domestic spaces, Tibetan wisdom pushes us to remember that *no* environment is exclusively human – and great care must always be taken.

Possession and Provocation

Of course, in some cases, illnesses caused by unseen beings are seen to take on a distinctly more *supernatural* flavor, including some cases that might be described as 'possession.'

In indigenous Himalayan traditions, oracles and spirit mediums have long played important roles in religious and social life, acting as key intermediaries between the human and more-than-human world. After the adoption of Buddhism, the practice was largely retained through the channeling of semi-enlightened protectors by established oracles. The Nechung Oracle, the most famous, acts as the official medium for Pehar, a key guardian figure tied to the establishment of Tibet's first Buddhist monastery. For many centuries the Tibetan government has relied on spirit mediums such as this to consult the wizened spirits of the Earth and solicit their input on major decisions.

In Tibetan Medicine, this kind of oracular possession is called *lha-dön*, or 'provocation from the gods.' To be touched by such an affliction is seen as both a blessing and a curse – at the onset, 'god possessions' can be quite dire, sometimes leading to grave outcomes, but if the patient can make it through the initial attack, then it's believed that they can learn to hone their skills and become a kind of oracle. The physical vitality of such individuals is believed to be negatively impacted by their craft, but *refusing the call* can equally have dire consequences.

Beyond *lha-dön*, other forms of spirit possession have even fewer redeemable qualities. The symptoms of such disorders are usually derived from the characteristics of the offending spirits themselves, causing the patient to suddenly behave in the manner of the spirit that has afflicted them. A *nāga* possession, for instance, may cause a patient to become uncharacteristically radiant, with penetrating, bloodshot eyes and a desire to wear the color red. They might flick their tongues around, somewhat like a snake, and may even feel compelled to lie face-down on their stomach. Since most *nāgas* are strict vegetarians, patients will also shun meat in favor of dairy and sweet foods. If a patient were to suddenly exhibit these symptoms out of nowhere, and if provocation was further demonstrated through pulse and urine analysis, then a doctor might diagnose them as being functionally 'possessed' by a *nāga*. Of course, diplomacy is highly important in Tibetan Medicine. In many cases, if spirit influence is suspected, the patient is never told of the real 'cause' of their affliction, to avoid causing undue distress.

Treatment of such cases remains a sensitive affair, requiring a combination of both physical and spiritual therapies. Special herbal formulas are often prescribed to expel the influences 'from the inside,' but it's traditionally important that the entity itself is unbounded from the patient through spiritual and ritual means. This often includes the use of special incantations, offerings, cleansing rites, meditations, and the performance of meritorious deeds. If a distinct spirit class can be identified, then special ritual offerings are arranged, based on the individual proclivities of the spirit in question.[20] Once the external influence has been satisfactorily placated, conventional therapies like herbal formulas are believed to be more efficacious.

As fascinating as this all is, it's important to understand that *very few* manifestations of 'mental illness' are thought to be caused by 'spirits' in the Tibetan medical tradition. *Sowa Rigpa* maintains a highly complex and nuanced view of the causes of mental suffering and it's understood that something like 'depression' or 'anxiety' can result from personal hardships, chronic pain, physiological imbalances, emotional trauma, eco-spiritual afflictions, and countless other factors, and in each case must be dealt with individually and holistically.

It's asserted that a clear mind, specifically one that has fully dealt with its own traumas and mental afflictions, is the best protection against the psychological destabilization of spirit possession and provocation. But this is only a minor factor in disorders with a physical microbial basis. You can't dodge pandemics by thinking positively, nor by spiritually bypassing 'fear.' We *should* be afraid of the ramifications of our destructive ecological practices. If we were a bit more afraid of global pandemics, then perhaps we would react with more suitable urgency.

A Wake-Up Call

According to the Tibetan medical paradigm, human health is directly linked to the health of those around us. Not only do we need to maintain a sense of balance in our own body, energy, and mind, but also in our more-than-human communities. The Tibetan theory of *provocation disease* highlights the critical importance of maintaining good ecological relationships, as well as the consequences that come from not doing so. Infectious diseases are direct consequences of our actions, but they are not so selective as imagined acts of 'divine retribution.' It's rarely the case that those who *cause* the most harm experience the gravest ramifications – instead, these are often reserved for the working class and those in developing nations, whose land and labor have been disproportionately exploited in the wealthy's pursuit of profit. Ultimately, we will all face the consequences in every nation on Earth.

Because we have so far failed to see our ecological crisis clearly, we are on track for a very dire prognosis. Beyond the emergence of new infectious diseases, the coming centuries will bring more and more catastrophic heat waves, droughts, rising sea levels, extreme weather conditions, desertification, famine, and the extinction of countless species. Our condition is bound to be chronic and painful, and very soon it will also become irreversibly terminal. Covid-19 gave us a glimpse of just one of the symptoms we can expect to see increase as we pass the point of no return, though many have failed to recognize this as a glaring indication of our disease's progression.

But there *is* still reason to be hopeful. Treatments *are* available. And while they cannot 'save the world' any more than antibiotics can make us immortal, they *can* help us to recover a sense of what it means to be human in a more-than-human world and to improve the quality of life for many kinds of beings. We can significantly limit our risk by mitigating the environmental trauma that we've already caused, and by putting systems in place to prevent further damage. We need to abandon our systems of exploitation, *make amends*, and adopt radically new lifestyles – based on relationships, not paradigms of control. We also, most pressingly, need to recognize the inseparability of human and ecological health. As Jonathan Coope writes, 'Any efforts which conceptualise or compartmentalise health and healthcare apart from, or away from, their ecological contexts are increasingly implausible.'[21]

But before any therapies can really be effective, we must first make amends for the suffering that we've inflicted upon the more-than-human world. It's only from this starting-point that we can begin to create more sustainable ways of being. To really make amends we must first acknowledge that there is *somebody* on the other side. We must come to terms with the realization that our actions have real, *experienced* consequences. Even just apologizing to a tree, the ocean, or a wild animal can be a healing experience, because we are meeting them on shared ground, in pursuit of authentic relationship.

Infectious disease is an excellent example of how ostensibly 'mythic' or 'folkloric' constructs like spirit paradigms can bring us closer to a true understanding of nature than empirical observation alone. Galenic doctors deemed it perfectly rational that foul odors were the cause of illness, given that they were sensorily observable, especially in places with poor sanitation. But until European scientists were able to *see* micro-organisms directly, it was considered absurd to think that tiny unseen beings could fly through the air and make people sick. And even once we could see them with our own eyes, it took us quite a while to embrace them as dynamic agents rather than simple clusters of animated molecules. Even today, we generally struggle to think of viruses and bacteria as information-processing organisms themselves, thus limiting our ability to comprehend the full range of factors that give rise to disease outbreaks.

The ostensibly 'mythic' approach taken by Tibetan medical systematizers was ultimately a robust model for thinking about pandemics as dynamic interpersonal and ecological occurrences.

While the administration of powerful anti-microbial substances was necessary to attack the 'unseen forces' infiltrating the body, attention was equally paid to preventing contagion, building immunity, and addressing any potential ecological, social, or behavioral causes. This rather robust approach to treating and preventing infectious disease was developed not from looking through microscopes, but as a rational conclusion based on the animistic assumption that non-human beings are an active part of our world.

Prognostication is difficult. The rapid expansion of ecological destruction is already causing painful consequences – from the *local* collapse of biological communities to *global* pandemics and climatic changes. We cannot 'fix this' with our current ways of thinking. Recovery is not a matter of finding some universal magic pill to cure us of the disease of 'climate change'. We're not dealing with an acute condition but a deep-seated *chronic* disease, rooted in our very orientation to the world. Our therapeutic process will, therefore, be varied and multifaceted. Our end goal mustn't simply be a 'cleaner' source of energy or a new carbon-sequestering techno-fix. After all, if clean energy is used to fuel exploitation, then it isn't really a 'solution' to our problem. Our therapeutic goal must instead be to learn (or rather *remember*) how to *care* – for each other, for the planet, and for all non-human beings.

PART IV
TREATMENT

7

LIBERATING THE UNSEEN

Going Beyond Gods and Demons

Of all the world's 'religious' traditions, Buddhism has always had a particular scientific appeal. Its distinctly *analytical* praxis has made it perennially compatible with many different fields of knowledge. According to the Buddha, a virtuous life is not typified by *belief* in gods, or their veneration, but *how we behave*, and the ultimate goal of the Buddhist path isn't eternal paradise among the gods, but a return to the non-conceptual state of pristine awareness. *Awakening* – or *nirvana* – is simply the state that remains when we 'wake up' and remove the veils that obscure our experience of reality.

While spirit work was never the core focus of the Buddha's path to liberation, maintaining healthy relationships with non-human beings has always been an important component of 'right engagement' in an interconnected world. For this reason, Buddhism has a rich tradition of rituals and stories involving unseen non-human beings, some of which may be quite useful in our treatment.

By establishing that non-human spirits can 'abide' in plants, forests, rivers, mountains, and all kinds of topographic phenomena, Buddhist cosmology left an open space for the possibility that *many* kinds of natural phenomena *may* be 'spirited,' and that unseen beings are an integral part of the natural world. This is, by definition, an *animistic* worldview.

But for many modern Buddhists, particularly in the West, so-called 'spirits' have been routinely repackaged as as either metaphorical representations or energetic manifestations of strictly *human* mental afflictions. Even non-human animals are sometimes characterized as a strictly one-dimensional personification of ignorance. This psychological turn was deeply influenced by interactions between Buddhism and psychoanalysis in the 20th century, but while figures like C.G. Jung derived significant inspiration from their own understandings of Buddhism, the animistic value of Buddhism in an ecologically challenged world has been sorely underexplored.

Indian Buddhist texts speak of 'eight classes of gods and spirits' that interacted with the historical Buddha, but as the movement spread along the Silk Roads, it encountered numerous new peoples, landscapes, gods, nature spirits, and ways of relating with nature. Buddhism's basic theoretical openness to the existence of unseen beings as *sentient beings*, rather than gods or demons, enabled it to integrate with a diverse range of spirit ecologies.

When Buddhism came to Tibet, it took root in cultural environments with their own animistic traditions reaching back thousands of years. Buddhist systematizers didn't seek to eradicate existing relationships with indigenous spirits, but to recontextualize their existence in terms of Buddhist cosmology. This multidisciplinary dialogue between formal Buddhist philosophy and indigenous animistic knowledge produced some fascinating and unique insights into the dynamics of the more-than-human world.

Sages and Seekers

Even before Plato delivered his symposiums, Indian thinkers were in the midst of their *own* philosophical revolution. Figures like Mahavira (founder of Jainism), Siddhartha Gautama (founder of Buddhism), and other *shramanas*, or 'seekers,' sought to transform notions of truth, meaning, worship, and ethics, and to challenge the oppressive social and spiritual hierarchies of Vedic Brahmanism. Rather than determining the best way to please the gods, figures like the Buddha focused their attention on the pursuit of insight, wisdom, and freedom from suffering.

To the Brahmanical societies in the west of India, the peoples of the Greater Maghada region (the homeland of the Buddha) were thought of as 'fierce, rough-spoken, touchy and violent' outliers beyond the boundaries of Vedic civilization.[1] While the Buddha's clan could also trace many of their cultural and linguistic traditions to Indo-Aryan societies (and thus also PIE traditions), they did *not* follow the Vedas, or a caste system, but adhered to their own traditional practices centered around the veneration of the sun, tree spirits, and serpents.[2]

But even in their own communities, the *shramanas* were both visionaries and disruptors. Most agreed that liberation came about as a result of behavior and ethical *conduct*, not the flawless performance of rituals or rites of worship. Mahavira asserted that *even the gods themselves* were just sentient beings caught in *samsara*, the cycle of birth and death, along with plants, animals, humans, and myriad unseen beings. By Jain standards, to cause harm to *any* of them is classifiable as an unethical act. Causing *as little suffering as possible*, even if it causes personal inconvenience, is a core tenet of the Jain path to liberation. As it states in the *Acaranga Sutra*:

> *As the nature of this [i.e., men] is to be born and to grow old, so is the nature of that [i.e., plants] to be born and to grow old; as this has reason, so that has reason; as this falls sick when cut, so that falls sick when cut; as this needs food, so that needs food; as this will decay, so that will decay... Knowing them, a wise man should not act sinfully toward plants, nor cause others to act so, nor allow others to act so. He who knows these causes of sin relating to plants is called a reward-knowing sage.* [3]

Mahavira advised that his followers maintain a diet of plant-based foods that were harvested non-violently – that is, without mortally harming any plants or animals in the process. While picking fruits and vegetables was generally deemed to be sufficiently non-invasive, digging up roots or otherwise killing a plant was deemed to be a breach of Jain ethics.

In addition to following these stringent dietary guidelines, Jain monastics often wear masks so as to avoid the inhalation of insects, and may sweep the ground ahead of them as they walk to avoid trampling on any small creatures. Expectations are less stringent for non-monastic practitioners, who are simply instructed to cause *as little suffering as possible* in pursuit of their own wellbeing and liberation. But with its austere practices and strict dietary restrictions, the Jain tradition has always had a reputation for asceticism.

A Middle Way

Siddhartha Gautama, the founder of Buddhism, was another key visionary in the *shramana* revolutions. By the time of his birth in India, these movements were already in full swing, with many people opting to renounce worldly life (and also socio-political hierarchies) in pursuit of spiritual liberation (*moksha*). To prevent his son from following in such reckless footsteps, Siddhartha's father, King Śuddhodana, sheltered him from all serious existential matters, including knowledge of suffering and death, for the first 29 years of his life. But his attempts to steer his son away from a path of renunciation ultimately backfired, and a fated encounter with a sick person, an elderly person, a corpse, and a seeker of truth turned Siddhartha's understanding of the world upside down, ultimately compelling him to leave his royal palace and young family in pursuit of answers to life's most important questions.

Siddhartha spent six years living with a band of wandering ascetics, studying under a guru who was likely himself a Jain master. It's said that he ate only one grain of rice per day in these years – itself a rather extreme form of dietary restriction based on concerns for vegetal suffering. But as he progressed on his path, he came to view this kind of asceticism as a neurotic and reactionary undertaking. While extreme luxury was clearly untenable as a path to spiritual realization, he doubted that self-inflicted misery was a suitable alternative. His 'middle-way' approach was, both philosophically and behaviorally, designed to foster a more neutral attitude between the extremes of Jain asceticism and Brahmanical laxity.

While he ultimately agreed with many of Mahavira's ideas, including the foundational principle of treating other beings nonviolently,[4] the two seekers disagreed on exactly *which* beings should be included in this scheme. Mahavira explicitly acknowledged that *plants* were sentient beings, but the Buddha apparently disagreed. While extant Buddhist teachings take a clear stance on the ethics surrounding our treatment of animals (and even unseen spirits), *plants* have been written out of the conversation altogether. It's almost certain that this was a direct result of the Buddhist dedication to *moderation*.

While many Buddhist societies today idealize vegetarianism, it has never itself been an essential feature of the tradition. For monastics, the Buddha established a tradition of begging for alms with utter dietetic indifference, thus bypassing any moral involvement in the agricultural process. For laypeople who could not simply beg for alms, he advised remaining as far down the supply chain as possible when eating animal products associated with slaughter. So long as a piece of meat had passed through *three hands* before reaching one's plate, and so long as the animal had not been slaughtered specifically for a practitioner or in their presence, his conclusion was that no karmic detriment could result from eating it.[5] It's clear that he was attempting to fashion a middle-way approach between rigid self-denial and uncritical indulgence. But while many Buddhists do themselves choose to adhere to a vegetarian or vegan diet, restricting the consumption of plant foods on the basis of vegetal ethics has never been a Buddhist practice.

In the monastic code of conduct (*Vinaya*), the Buddha does, however, repeatedly speak out against cutting down trees or causing needless harm to plants, particularly if they are understood to be *spirited*.[6] But once a tree *spirit* has been relocated, it is then considered morally inconsequential to destroy the tree. Given what we now know about the observable sentience of plants, including their potential to experience suffering, this idea warrants serious revision. But in practice, we, as animals, must rely on the bodies of plants for food and numerous other resources. Rather than bypassing this ethical quandary by deciding that *because it's too difficult, they must not be conscious*, we could alternatively just *accept* that most agriculture produces some degree of non-human suffering. If we could simply hold this as a basic truth, we might be naturally compelled to act more carefully and conscientiously, to waste less, and to try to support the welfare of plants rather than simply exploit them as resources.

Due to capitalism and industrial farming, things are obviously far more complicated today than they were 2,500 years ago, or even 500 years ago. Capitalism gives us no shortage of opportunities to dissociate ourselves from the abuses of industrial exploitation and the degradation of non-human environments in the name of profit. It's easy to stay *three steps away* from the atrocities of slaughterhouses, sweatshops, and diamond mines. But it's still our *money* that keeps the engine running, and this kind of spiritual bypassing is no reasonable way to avoid moral responsibility.

The Tree of Awakening

It's notable that not all Buddhist traditions were so dedicated to plant blindness. Early Japanese masters in the Zen, Tendai, and Shingon schools openly questioned whether plants were sentient beings, and thus capable of attaining liberation. Tendai master Ryōgen (912–85CE) argued that the 'phases' of a plant's life were themselves analogous to the stages of the spiritual path, concluding that 'we must, therefore, regard these [plants] as belonging to the classification of sentient beings.'[7] Chujin (1065–1138CE), another Tendai master, further argued that it was unnecessary to rely on anthropomorphic tropes when conceiving of vegetal 'Buddhahood,' writing:

> *As for trees and plants, there is no need for them to have or to show the thirty-two marks [of Buddhahood]; in their present form – that is by having roots, stems, branches, and leaves – each in its own way has Buddhahood.*[8]

Naturally, a hypothetical 'enlightened plant' wouldn't suddenly sprout a crown protuberance and sit in the lotus posture like a human being. But in Buddhism, as already mentioned, what is termed 'liberation' is fundamentally just a return to our natural state – there is nothing inherently 'human' about it. And, despite its occasional plant blindness, Buddhism has never been overly anthropocentric. The acceptance of nature spirits offered a great deal of flexibility for the inclusion of non-human and non-animal beings, and it should really come as no surprise that plants should still hold a degree of venerability in the Buddhist tradition. After all, the tradition itself arose out of a long sit under a shady tree.

After leaving asceticism in pursuit of a more moderate path, Siddhartha ventured southward and eventually came upon a mighty tree.[9] In line with his ancestral customs, he prostrated himself seven times before the tree, and made himself a seat of *kusha* grass on the eastern side of their trunk.[10] He vowed not to move until he had attained complete awakening, just as Mahavira had done under his *ashoka* tree in the past. Even if asceticism was not the path for him, he was still in pursuit of a similar state of realization.

Siddhartha's tree was a member of the *Ficus religiosa* species, the 'sacred ficus,' a particularly long-lived plant which was already well-established as a holy tree species in India, being revered in the *Rig Veda* as the divine *peepal* tree, and in later Hindu scriptures as the mystical *aśvattha*, 'king of trees,' both of which have important spiritual associations.[11] Often perceived as an earthly embodiment or 'abode' of gods, *yakṣa* spirits, and other powerful beings, these trees were frequently worshipped, ritually circumambulated, and propitiated for blessings.

After meditating under the Bodhi Tree for 49 days, it's said that Siddhartha came face to face with Māra himself – the very embodiment of egocentricity, and master of all the forces that are antithetical to awakening. Māra tempted Siddhartha with sensual pleasures, threatened him with violence, and ultimately attempted to undermine his confidence and worthiness of attaining awakening, taunting, 'Who are *you* to attain enlightenment? You don't even have a teacher to bear witness to your awakening!'

At this decisive moment, as dawn began to break, Siddhartha touched his hand to the soil beneath him and said, 'The *Earth* is my witness,' and the ground quaked with Mother Earth's testimony.

Just then, as the rays of the sun crept across the horizon, Siddhartha attained enlightenment. He was now *the Buddha*.

This event is often commemorated in Buddhist iconography with the classic motif of the Buddha touching the earth. In some more ornate depictions, a goddess can even be seen reaching up from the chthonic space below – Mother Earth herself, *Prithvī Mātā*, bearing active witness to the Buddha's liberation.

Even in the Buddha's own lifetime, devotees would travel from far away to honor the sacred Bodhi Tree, bathing their exposed roots with perfumed water and milk.[12] Cuttings of the tree were sometimes planted to inaugurate new temples and monasteries – the most famous of which went to Sri Lanka in the 3rd century BCE and remains the world's oldest cultivated tree with an established written record.[13] Within a few centuries, the site of the Bodhi Tree became a bustling pilgrimage destination, complete with an elaborate temple complex, commissioned by Emperor Ashoka in the 3rd century BCE to honor this new *axis mundi* of the Buddhist world.

According to legend, Ashoka's famed obsession with the tree led his queen to mutilate them with poisonous thorns in a fit of jealousy. The tree regenerated, only to be cut down a century later due to political shifts in the region – a process that was repeated many times over the course of the tree's many lives, for in addition to being one of the world's most *spiritually* significant trees, the Bodhi Tree is also perhaps the most politically embroiled. Upon inspection in 1862, British archeologist Alexander Cunningham noted that there seemed to be *many* different 'trees' clustered into a massive composite entity, a result of ongoing cycles of destruction and regeneration in an ever-shifting human social climate. Today's tree was only planted 140 years ago, by Cunningham himself, after the predecessor was killed in a violent storm.[14]

The iconographic and ritual significance of the Bodhi Tree has somewhat diminished over time, particularly in regions where *Ficus religiosa* doesn't easily grow (like Tibet), but in early Buddhist art the Bodhi Tree was wholly synonymous with awakening. Until around the 2nd century CE, no depictions of the Buddha in human form were used in Buddhist art, so the tree frequently took his place, usually flanked by 'tree spirits' or reverent human admirers. It wasn't until the Gandharan artistic movements in the early centuries of the Common Era that iconographic depictions of the Buddha himself began to emerge and the Bodhi Tree faded into the background.

While even to this day, the distinctive heart-shaped leaves of the Bodhi Tree can be found worldwide as symbols of Buddhism, the strictly symbolic value of the motif has gradually eclipsed appreciation for the 'personality' of the living tree. As Rhys Davids mused in the *Life of Gautama* (1870), 'The Buddhists look upon the Bo Tree as most Christians look upon the Cross.'[15] But in the Buddha's own social environment, the tree was an important and spirited character on his journey to awakening. In quite the reversal of the Abrahamic Tree of Knowledge, the Bodhi Tree offered a means to existential *freedom* rather than existential *bondage*. In providing a support for Siddhartha's awakening, she facilitated his return to *nature* in a mundane and absolute sense.

Trees at the Beginning – Trees at the End

While the Bodhi Tree is certainly the most famous tree in Buddhism, Siddhartha's stint under the Bodhi tree wasn't the first or last meaningful tree-encounter that the Buddha had. It's said that his mother, Mayadevi, went into labor amidst a grove of Sal trees (*Shorea robusta*) near her village in modern-day Lumbini, Nepal. As the birth grew nearer, the trees lowered their branches to support the queen as she delivered baby Siddhartha standing up! As the story goes, the newborn Buddha-to-be immediately took seven steps forward, with a lotus flower blossoming in every step.

We're told that Mayadevi passed away a week after Siddhartha's birth (undoubtedly due to complications in childbirth) and was reborn in Indra's kingdom of the gods atop Mt Meru, where, in her new divine form, she remained an engaged maternal figure in the Buddha's spiritual life.[16] But in his later years, he began to worry that she would be stuck in *samsara* if he were to pass away and 'leave' the world entirely. To help him see her one last time, the mountains are said to have lowered their peaks to lift him up into the heavenly realms, where he was reunited with his mother amid a grove of divine trees, sometimes said to include the night jasmine, *Nyctanthus arbortristis*. In this grove of the gods, the Buddha offered his mother and her friends some final teachings before returning to the terrestrial realm and completing his life's work.[17]

When he eventually passed away at the age of 80 in Kushinagar, an assembly of Sal trees were present at his deathbed, sharing in the grief of his disciples by releasing an 'abundance of untimely blossoms, which fell upon the Tathāgata's body, sprinkling it and covering it in homage.'[18] This rather touching motif is depicted in numerous representations from the Gandharan period, with tree spirits often shown weeping in the branches above the deathbed.

While these plant-centered elements of the Buddha's story are often underplayed in accounts of his life, it's worth acknowledging that his birth, enlightenment, and death were all witnessed by responsive and sensitive plants. Given that the society he was born into had long identified with sacred trees, practiced sun worship, and established relationships with more-than-human beings, this is much to be expected.[19] What really changed with

Buddhism was the understanding of *what* non-human beings really *were*. Rather than being objects of worship or gods with certain powers, they were seen as simply *sentient beings* – sensitive creatures in an array of diverse forms, all of whom were caught up in their own veiled and impermanent experience of life.

The Wheel of Existence

While some Indian traditions viewed reincarnation as an ongoing process of existential *promotion* based on the dutiful performance of social and ritual obligations, the Buddha took a very different approach. Rather than climbing an existential ladder to divinity, he argued that the conditions that determined our future incarnations were entirely dependent upon our present actions (i.e. *karma*). We are constantly sowing the seeds of our future experiences, but the ultimate goal of Buddhism is not to be reborn in some heavenly realm among the gods, but to *break free* from the cycle of rebirth altogether and return to our natural state of unborn awareness.

What is the best way to do this? In a Buddhist worldview, the only acts that can be classified as 'unethical' are acts that cause other beings to suffer (especially needlessly and/or against their will), and the best way to stop suffering is to replace animosity and complacency with *active compassion*. It is *empathy*, not rote adherence to rules, that changes hearts and undermines *samsara*.

There are *infinite* sentient beings to be found in the Buddhist cosmos, inhabiting infinite world systems that arise and decay perpetually across beginningless time. The foundational *consciousness* that underlies each living being is believed to have been cycling through *samsara* all this time, leading to the traditional belief that 'every sentient being has been our mother in a previous life.'

In the Tibetan tradition, it's usually taught that there are *six* specific *realms* into which an individual can be 'born' – the realms of the gods, demi-gods, humans, animals, and hungry ghosts, and the frightful hell realms. In every realm, the rules are basically the same: We arise, we live, we suffer, we die, and then the kinetic momentum of our karma propels us into a new birth.

While a 'precious human birth' is often described as the best kind of incarnation for cultivating the potential for awakening, many Buddhist texts speak of the spiritual capacities of non-humans, and it's asserted that 'extraordinary bodhisattva discipline is also open to gods, *nāgas*, and other non-human beings.'[22] There is no form of sentient being (human included) that can be characterized as wholly good or wholly evil. We are all morally ambiguous – capable of both benefit and harm – and equally capable of waking up from our samsaric dream.

Compassion for the Gods

The Buddha lived in a world in which *devas*, *asuras*, and disembodied spirits were widely understood to be 'real.' Rather than try to undermine belief in their existence, the Buddha simply sought to reform attitudes *toward* these beings. So, while 'gods' are not treated as rulers or saviors, they are very much a part of the Buddhist cosmos. Most are seen as being far removed from our own planetary affairs, while others are thought to frequently involve themselves in the lives of human beings. But either way, in the Buddha's view, even the gods are in fact trapped in *samsara*.

Worse, it's said that the relative privilege and bliss of their experience prevents them from ever seeking the true nature of reality or cultivating virtue in their long years of life. For this reason, even the highest gods can fall into terrible rebirths in future lives as a result of their karmic complacency. Kangyur Rinpoche writes:

> *Yet, when the propelling force of his karma is spent, even Indra himself will fall... As a sign of [the gods'] approaching death, their robes grow foul and rank; they find no comfort on the couches; their garlands wilt; and they are abandoned by the goddesses who had previously attended them. And drops of sweat, which until then they had never known, now break out upon their brows and limbs. Gone are the divine maidens who used to bathe them. They are beside themselves with dread... their friends and family can only call to them from afar, sending them wishes that they might be born again among the gods or in the human realm. And as they gaze upon their friends, they weep in terrible grief. For seven days, according to the reckoning of the heavenly realm, they experience the agony of death and transmigration – a depth of suffering utterly unknown in the human state.*[20]

This is quite a different approach to 'gods' than we find in other religious traditions, including in ancient Vedic religion. In Buddhism, these gods may very well 'exist,' but they are understood to be distinctly mortal, imperfect, and no more 'in control' of the universe than we are. Even Brahma and Indra are merely *other sentient beings* stuck in their own version of reality, and according to the Buddha, the very best way to deal with them is to treat them with compassion.

The Buddhist *asuras* (often glossed as 'demi-gods') are, like their Hindu counterparts, perceived as chaotic anti-gods who are consumed with envy of the riches and glory of the *devas*. This compels them to wage fruitless wars against the gods for aeons on end.[21] In principle, the archetype of the *asura* in Buddhism largely functions as a warning against envy and militancy, but in practice, *asuras* also frequently appear as gatekeepers of the underworld and guardians of treasure, usually making their homes in caves and enchanted places.

The realm of *hungry ghosts*, or *pretas*, is another common domain of Buddhist spirit beings. In traditional sources, they are usually described as miserable, starving beings with 'mouths as small as the eye of a needle, while their bellies are the size of an entire country.'[22] Their inability to effectively nourish themselves is presented as a karmic consequence of insatiable desire. But many Tibetan theorists speak of a *second* class of *pretas* who are distinctly less miserable and more dynamic, including many so-called Tibetan 'nature spirits.' While approaches to cosmological categorizations can differ, the general Buddhist consensus has always been that *unseen beings* are an active and vital part of the natural world.

Spirit Classes

When it comes to Buddhist (and broader Indian) nature spirits, we must of course begin with the *nāgas*. The Sanskrit word *nāga* means 'snake,' and indeed these beings are often seen as serpent- or dragon-like in form. But they can also present themselves in humanoid or semi-humanoid form, even taking on rather nymph-like (or even elf-like) characteristics.[23] *Nāgas* can be perceived as wizened guardians of secret knowledge, or as bearers of disease.[24] On some occasions, they're even said to have interbred with humans, leading to the birth of great heroes like *King Gesar*.[25]

Nāgas are closely connected with the underworld, land, and bodies of water. Also, like the nymphs of ancient Greece,[26] they are often closely related to trees and vegetation. They are not *synonymous* with the plants themselves, but *coeval* with them, and as their lives are intertwined, to damage the vegetal or topographic *abode* of such a spirit is to bring harm upon them. Thus it's always advised that we approach 'spirited' places with the utmost respect. If we bring harm to beings like the *nāgas*, environmental and pathogenic crises can ensue. But if we maintain a healthy and balanced relationship with them, we can experience benefits. Many forms of Tibetan ritual magic, including the summoning of rain, are facilitated in part through engagement with *nāgas*, making them key spirit intermediaries between the human and more-than-human world.[27]

When it comes to 'cosmological' classifications, *nāgas* are a great example of just how *slippery* spirit paradigms can be, especially when transmitted and translated across cultures. Buddhist sources are quite conflicted on how to categorize *nāgas* – some identify them as animals, while others describe them as a class of *pretas*, *asuras*, or *devas* bound to our physical world. But the *nāga* lifestyle is distinctly *human* in many respects, with human-like class structures and a particular interest in material wealth.[28] As such, it's difficult to pin down a precise identity for the *nāga* in Buddhism.

While there are many stories about *nāgas* who abide in trees,[29] *yakṣas* remain the most prominent 'tree spirits' in the Indian Buddhist world. An entire chapter of the *Samyutta Nikaya* sutras is dedicated to the Buddha's discourses on (and *to*) these beings, who are often seen to inhabit sacred trees and groves. In some instances, he offers them teachings, while in other cases it's the *yakṣas* who bestow teachings on others.[30] It's easy to see a distinct Buddhist affinity for such 'tree spirits,' and a generally positive view of non-human beings, particularly in their capacity for virtue. Even if plants were rejected as sentient beings for practical purposes, paradigms such as these left open an important space for enchanted engagement with plant *spirits*.

Active Engagement

Nature spirits like *nāgas* and *yakṣas* remain active players in many Buddhist ritual traditions. Their roles expanded considerably from around the 4th century CE, when tantric modes of spiritual practice began to emerge from the fringes of Indian Brahmanical society. This led to the co-emergence of 'Buddhist' and 'Hindu' streams of tantra, but many prominent tantric rituals, deities, and texts flowed effortlessly between traditions.

Hindu practitioners use tantra to enter into union with 'divine consciousness,' while Buddhists seek to return to the 'unborn nature of mind,' but both traditions seek to accomplish these aims using similar methodologies. The world of the tantric practitioner is one that is absolutely teeming with magic, miracles, and powerful supernatural forces. Instead of relying on quiet Buddhist contemplation *or* Hindu paradigms of worship to sow the seeds of liberation for harvesting at some point in the future, tantra seeks to establish the enlightened state in this very life. It makes use of countless complex visualizations, secret mantras, advanced yogic techniques, and a diverse array of rituals to engender an embodied *experience* of enlightenment. Habituation to these processes, often through years of repetition, is believed to gradually undermine our mind's tendency to produce an experience of reality based on suffering and replace it with an experience of awakening.[31]

For a tantric practitioner, *all* phenomena, both 'good' and 'bad,' 'pure' and 'impure,' are meant to be perceptibly transformed into the miraculous manifestations of an enlightened mind. Even terrifying and haunted charnel grounds are imbued with vast spiritual power, as are the beings who inhabit such enchanted locations. Many of the most beloved tantric deities are indeed enlightened demons who reside in desolate cemeteries.

Through tantra, 'nature spirits' were able to emerge from the background and become central characters on the spiritual path. Some were regarded as allies and protectors, while others were subjugated and compelled into service through magical operations. While the ethics surrounding the subjugation of non-humans are questionable, by 'controlling' unseen beings in this way, an accomplished *tantrika* can presumably change the weather, banish unfriendly spirits, heal incurable diseases, fly through space, and perform an array of other miraculous feats. But in some cases, tantric magic is invoked to subjugate and even destroy non-human entities.[32] Rituals of banishment and annihilation are an integral part of the classic Indian tantric arsenal, and many of the most important meditative and ritual procedures are modeled after the wrathful 'liberation' of such beings. While the more-than-human ethics of such models were debated by some Tibetan systematizers, as we'll see below, the underlying principle of spirit collaboration remains an important one in Tibetan Buddhist tantra.

While tantra itself is an esoteric or 'secret' movement, there are certain basic principles that are universally applicable and the archetypes and paradigms that we identify with can deeply impact how we experience 'self' and 'environment.' We don't need to settle for misery, alienation, or transcendent self-denial. We don't need to reject nature in order to experience awakening. We don't have to stop singing and dancing to discover philosophical depth. We don't need to reject love to realize non-attachment. This life is right here, right now – we might as well make the absolute most of it.

Conversion and Consent

In the Buddhist worldview, there are *no* sentient beings, including the king of the gods himself, who can effectively *save* us from our own existential misery. Other beings might offer the counsel of a teacher or friend, but worshipping them isn't going to do much in the grand scheme of things.

On this basis, indigenous gods, monsters, and nature spirits didn't need to be purged from a cultural tradition for Buddhism to effectively take root. In all the places that Buddhism was adopted, it was actively and dynamically *integrated* with indigenous systems of knowledge.

Many indigenous gods and spirits were assimilated into the Tibetan Buddhist cosmos. Some were absorbed as mischievous demons, others as more neutral and ambiguous spirits connected to the natural world. But some pre-Buddhist Tibetan gods were formally admitted into the Buddhist pantheon as guardians and protectors, many equipped with their own dramatic conversion legends. Most of these accounts revolve around the legendary tantric wizard Padmasambhava,[33] who is said to have come to Tibet in the latter half of the 8th century at the request of King Trisong Detsen and used his magic to subdue and *convert* the local spirits, binding many of them under oath as protectors of the teachings.

The idea that indigenous spirits need to be exorcised of their 'pagan' ways before being forcefully assimilated into a new religious orthodoxy is strikingly familiar, and potentially problematic. But fortunately this was not the *only* story that was told in the Tibetan world, and approaches to dealing with nature spirits in the Tibetan Buddhist tradition are actually quite diverse. Ritual experts are often consulted to gauge the condition of local spirits before engaging in invasive behaviors, like felling trees or breaking ground for a new building.[34] Such acts are conventionally preceded by the dedication of offerings and a formal request for 'permission,' reinforcing social perceptions of an *exchange between persons* rather than human-centered exploitation.

Even when destructive acts are performed out of necessity and noble intent, there remains an obligation to acknowledge their impact on the environment and to mitigate potential suffering by seeking some form of 'active consent.'

A Gracious Approach

When it comes to 'gods' and 'demons' in a more classical sense, Buddhism asserts that they aren't something we can find 'out there,' but rather phenomena that can arise in the space of the mind. This is a fairly enigmatic feature of Buddhism as far as 'religions' are concerned, and is one of the things that first drew me to the tradition 20 years ago.

In Tibet, the most famous teachings on the nature of gods and demons are attributed to Machig Labdron (1055–1154CE), matriarch of the *Chöd* (Wyl. *gCod*, 'Severance') tradition. *Chöd* is one of the only Tibetan Buddhist lineages *not* to lay claim to an original Indian source,[35] and it's the only major Tibetan Buddhist tradition to have been founded by a historical woman. Over time, Machig's teachings gained widespread popularity, making their way into nearly all major Tibetan Buddhist schools.

While *Chöd* was once a comprehensive practice lineage with many different components, today the term is mainly used to refer to a ritual known as the *Lüjin* – an iconic rite famed for its rich iconography and musical atmosphere, with haunting ancient melodies accompanied by a drum, bell, and thighbone trumpet. It is performed in frightening and wild places, which serves numerous purposes, including engendering a necessary degree of altruistic bravery in the face of the things that scare us, and also a degree of active engagement with the powerful spirits of the world.

When Machig was in her twenties, she received a tantric initiation with a group of her peers from a teacher named Sönam Lama. During the ritual, it's said that she stood up, disrobed, and floated into the air, passing through the walls of the temple, and flew into the branches of a mighty tree. There, she became aware of the presence of a fearsome *nāga* king. Troubled by the young woman's presence, the *nāga* threatened to attack her with his army of angry spirits. But instead of cowering in fear, she responded with pity and generosity, offering her body as food for the *nāga* in an act of selfless love. Disarmed by her compassion, he decided instead to commit himself to the protection of her lineage.[36]

This legendary account stands in rather stark opposition to the stories of Padmasambhava. While *Chöd* is often spoken of as a dark and macabre practice, Machig's gracious approach to dealing with frightening spirits is in fact a profound exercise in radical compassion for non-human beings.

In the course of a standard *Chöd* rite, the practitioner 'ejects' their consciousness from their body, transforming into a powerful female deity, while their physical remains are imagined to be alchemically transformed into a boundless cornucopia of magical offerings. The practitioner then invites a vast host of seen and unseen beings to partake of the feast, which is offered in turn to enlightened beings, protectors, and all the sentient beings in the universe. Concepts of 'inside' and 'outside' are radically upended, with the most contemptable and frightening spirits summoned to the front of the line as venerable *guests of honor*. Malevolent forces and even spiritual vectors of disease are thus pacified through maternal love and generosity, not through acts of ritual violence and banishment.[37]

In Tibet and beyond, *Chöd* is often associated with the concept of *devils/demons* or *māra* (*düd* in Tibetan, Wyl. *bdud*), but Machig states, 'That which is called "devil" is not some actual great big black thing that scares and petrifies whomever sees it. A devil is anything that obstructs the achievement of freedom.'[38]

The psychological and existential applications of this insight are aptly unpacked in Lama Tsultrim Allione's 2008 book *Feeding Your Demons*,[39] where she writes:

> *Our demons are not ancient gargoyles from eleventh-century Tibet. They are our present preoccupations, the issues in our lives blocking our experience of freedom... Demons are ultimately part of the mind*

and, as such, have no independent existence. Nonetheless, we engage with them as though they were real… In the end, all demons are rooted in our tendency to create polarization. By understanding how to work with this tendency to try to dominate the perceived enemy and to see things as either/or, we free ourselves from demons by eliminating their very source [i.e. dualistic egocentricity].[40]

This is a profoundly pertinent lesson with distinct relevancy in the modern world. Real gods and demons aren't somewhere *out there*, but in the space of our own minds. In their most basic manifestations, *gods* are simply the things or beings that we perceive as 'helpful,' while *demons* are the things or beings that we perceive as 'harmful.' A spouse can act as a 'god' in our lives, but if they betray us, then they can quite quickly transform into a 'demon.' Later, if we find an even *better* spouse, we may recast our flawed ex as a god or 'blessing' in disguise. This is, at the end of the day, a common dynamic in our perception of 'gods' and 'demons' in our lives, including the supernatural kind.

Machig warns against becoming attached to 'worldly gods,' even powerful spirits, since obsessing over the things that seem to help us ultimately just turns them into demons. This has clear implications for our psychological experiences, but a more literal interpretation is equally valuable in a world full of dangerous gods. Belief in the existence of 'gods' is not itself a cause of suffering, but our attachments to them can have dire ramifications, not least of which is the incessant violence perpetrated by religious zealots and extremists. What's more, our projection of divinity onto external sentient beings robs them of their own vulnerability, thus preventing us from being able to relate to them from a place of genuine care.

While the psychological and philosophical approach to Machig's teachings has become standard in western approaches to *Chöd*, there is another side to her argument that often remains unacknowledged. If the wild and powerful spirits of the Tibetan landscape are not, in fact, *demons*, what are they? According to Machig, the answer is deceptively simple: they are simply *other sentient beings* caught in their own mode of temporary existence. Sentient beings can't truly be 'gods' and 'demons' in an absolute sense because they are neither omni-benevolent divinities nor omni-malevolent demons, and they ultimately have no fundamental power over our fate.

As followers of the Buddha, Machig reminds her students that these sentient beings are the very embodiments of our own past mothers, pushing us to cultivate an organic sense of love and compassion toward even the most frightening ghouls.

Cutting Through Attachment to the Self

Machig's approach of *feeding* our demons, rather than *fighting* them, is profoundly valuable on many levels, including our relationship with our own psyche. When we alienate ourselves from our own 'shadow' elements, we only end up empowering the obstructive capacities of our inner 'demons.' It's only through acknowledging the underlying *needs* they obscure that we can resolve our inner conflict.

Machig did not view unseen beings as mere allegories for our inner struggles, however. Her Tibet was a land of innumerable unseen beings, and she accepted them as non-human residents of a still-enchanted world and asked her students neither to reject them nor to worship them, but simply to acknowledge their existence as sensitive beings in need of love.

By transcending our fear (or awe) of the otherworldly 'other' and establishing an authentic sense of kinship, we can simultaneously pacify our 'inner demons' and repair our broken relationship with the more-than-human world. On a most essential level, however, *Chöd* is a process of severing our 'core demon' of egocentricity – our basic attachment to the notion of a fixed 'self.'

According to Buddhist philosophy, a 'self' is an impermanent and composite phenomenon. Even our bodies are not truly 'one being' – as already noted, we are holobionts, entire microcosms of human and non-human cells working together to sustain our collective life. Zoom out a bit more and you'll find that our own seemingly autonomous bodies are both sustained by our 'environment' and constantly contributing to it, with our biological waste and outgases contributing to the lives of other beings and the perpetual terraforming of our planet. There is no *autonomy* in nature – 'we' are an infinite array of interrelated agents, and by relinquishing our desperate attachment to a self-centered universe, we can come to truly care for *all* sentient beings. Given that all of our 'master identity' complexes are fundamentally rooted in egocentricity, any disruption of our self-centered worldviews can have a significant impact on the societies we help to constitute.

Machig's tradition leaves us with some of the most lucid and profound 'Buddhist' teachings for relating with unseen beings. The ethic of *feeding not fighting* can radically alter the way we approach not only our own psychology, but also our ecological and social paradigms. Even Covid-19 can be approached in this light – rather than viewing a pandemic as an enemy that must be 'fought,' we *could* think of it as a painful symptom of an environment that is being mistreated and exploited. We need to approach such circumstances with compassion and care, rather than aggression, apathy, or machismo. Even the virus itself is not a 'demon,' but rather another kind of 'being' trying to survive. While it's imperative that we prevent illness and loss of life and embrace our mutual responsibilities to one another as members of a *society*, we also need to take a serious look at the underlying ecological causes of crises like pandemics. Situations like Covid are symptomatic of our lack of *care*, both for one another and for the non-human beings with whom we share this world.

Whether or not you are particularly interested in Buddhist cosmology, engagement with ideas like these allows us to imagine the so-called 'natural world' in an entirely different light. It's worth noting that the use of the word 'nature' to talk broadly about 'non-human environments' is a rather enigmatic idea with very few correlates outside of European cultures. The very notion that there exists some *external* 'nature' that is fundamentally sundered from 'humanity' is merely a story – a *myth*. But there are countless other myths out there, many of which cut far closer to what we know to be *true*. These voices need to be heard.

8

ENTERING THE PERILOUS REALM

The Revolutionary
Power of Myth

We are going to need more than good data to meet the challenges posed by the Anthropocene – we're going to need good stories, and new ways of imagining our place in the world. Every great revolution in human history has been accompanied by new myths. But we must find a way to engage with this essential tool in a way that avoids patterns of dogmatism, exclusivism, and exploitation. New myths can help us find new paths forward, but they can also lead us deeper into delusion. For this reason, the realm of myth is a perilous and dangerous place, but the path to recovery leads directly through it.

What is a Myth?

We've already encountered many types of myths in the course of this book. I've intentionally been a bit loose with the term, in part to illustrate the fact that *most* of our lives are ultimately built on myth – both the mundane and supramundane varieties. But 'myth' is a notoriously complicated word. Sometimes it's used pejoratively to refer to a mere falsehood, while other times it's used to identify a 'story about something significant,'[1] especially in old cultural traditions. But, as Joseph Campbell states in *The Hero with a Thousand Faces*:

> *It wouldn't be too much to say that myth is the secret opening through which the inexhaustible energies of the cosmos pour into human cultural manifestation. Religion, philosophies, arts, the social forms of primitive and historic man, prime discoveries in science and technology, the very dreams that blister sleep, boil up from the basic, magic ring of myth.*[2]

In the broadest of strokes, myths are stories which are usually constructed to illustrate important 'truths.' *Historical* myths speak to the affairs of the past, often embellishing them with cosmic grandiosity; *etiological* myths establish the causes of things, explaining why things are the way they are; and *psychological* myths seek to narrativize our internal human experience.[3] Another important kind of myth is the *charter* myth, relied upon for the establishment of a new religion, institution, or state, which weaves together many layers of myth to establish a convincing paradigm of authority.

We all go through life being conditioned by myths – it's simply a matter of whether we acknowledge it or not. The stories we tell have a profound impact on our perception of the world, helping us to formulate and contextualize our identity, our values, our concept of justice, our hopes and fears, even our ideas about the very meaning of existence.

We are, of course, quite familiar with the myths that use fantastical motifs to legitimize the authority of a human establishment – such myths form the basis of many of the world's religious traditions. But myths needn't be tied to divine covenants or powerful institutions to have real-world impact. Some emerge from a far deeper and more earnest space, arising at the threshold of the more-than-human world. We might encounter such myths in the inherited stories of our ancestors, or perhaps even in a work of literature. Whatever the source, when stories such as these really *speak* to us on a primal level, we would do well to listen. They may just lead us into a new paradigm.

We are all mutable, responsive, and imaginative by nature – capable of creating entirely new lived realities simply by collectively imagining them. As Yuval Noah Harari states, 'Humans control the world because they can cooperate better than any other animal, and they can cooperate so well because they believe in fictions. Poets, painters and playwrights are therefore at least as important as soldiers and engineers.'[4]

Harari specifically points out the importance of *science fiction* as an artistic tool for the technocratic age:

> *In the early twenty-first century, perhaps the most important artistic genre is science fiction. Very few people read the latest articles in the fields of machine learning or genetic engineering. Instead, movies such as The Matrix and Her and TV series such as Westworld and Black Mirror shape how people understand the most important technological, social and economic developments of our time. This also means that science fiction needs to be far more responsible in the way it depicts scientific realities, otherwise it might imbue people with the wrong ideas or focus their attention on the wrong problems.*[5]

Science fiction offers a kind of sandbox in which we can 'act out' some of the possible outcomes of the technological advances that inevitably lie ahead of us, giving us a medium for grappling with some of their psychological, sociological, environmental, and even political ramifications. Some science fiction may even approach the realm of the 'mythic,' but at least in its usual iterations, science fiction has some glaring limitations. Save for the occasional extra-terrestrial, which is generally either humanoid or monstrous, science fiction often falls into the worn-out tracks of anthropocentrism, usually rehashing some version of a 'mind over matter' trope to glorify the sovereignty of humanity.[6]

While stories such as these can still be quite useful for helping us make sense of the technological futures that lie ahead of us (including our relationship with artificial intelligence), they do little for reformulating our relationship with non-human beings in our own natural environment. For this, *other* types of story need to be told. If we want to improve and enliven our relationship with nature and come to a deeper sense of our own humanity, we will need to venture into the perilous realm of myth.

Falsehood and Fantasy

In modern times, many of our myths have lost a great deal of their enchantment. We don't imagine that George Washington sailed across the Hudson on the back of a talking whale, though we do think that Thanksgiving was a friendly meal shared between friendly Pilgrims and Native Americans. Myths such as these are not 'important stories' that enchant us with the truth, but rather simple falsehoods that delude us. Modern capitalism is built upon such falsehoods. The basic premise of consumerism is that 'more possessions will lead to greater satisfaction,' a *mythic* motif that has been repeatedly fed to us by the very corporations that benefit from our consumption.

But beyond the pragmatic use of falsehoods to manipulate society, myths can help us see our ordinary world in a new and enchanted light. A good story can transform even a rock into a source of limitless awe. But this mode of storytelling has largely fallen from its lofty position at the center of human society and been relegated to the oft-mocked domain of 'fantasy.'

As ridiculous and uncomfortable as it may initially seem, a healthier relationship with fantasy will prove to be a crucial component of our path to recovery. As Ursula K. Le Guin once wrote, 'What fantasy does that the realistic novel generally cannot do is include the nonhuman as essential... To include anything on equal footing with the human, as equal in importance, is to abandon realism.'[7]

While so-called 'realistic fiction' compels us to move ever deeper into *human* affairs, fantasy allows us to stretch beyond the limits of anthropocentricity and consider the experiences of the more-than-human world.[8] Much like science fiction in a burgeoning technocratic age, it offers us a kind of 'sandbox' in which we can psychologically and socially experiment with our recovery of natural enchantment and allow ourselves to explore new ways of engaging, including emotionally and morally, with non-human beings. Perhaps most importantly, it can facilitate this experience without falling into dogma or systems of rigid 'belief.' Fantasy remains a free space – a place where we can collectively taste the wonders of enchantment without the burdens of fear, shame, and guilt.

A really good story can be a lifelong companion. Long before I became absorbed in the Buddhist tradition, sparks of mythic enchantment were ignited by the works of J.R.R. Tolkien. I often speak of his works as my adopted mythology, much as Buddhism is my adopted religion, and I will use this for illustrative purposes throughout this chapter. It is a modern mythology, but modern stories are no less viable than their ancient counterparts in guiding us to an experience of 'truth.'

There is a saying in Tibetan, *jikshé kundröl,* or 'knowing *one* liberates *all*,' meaning that if we can truly comprehend *one* thing in its entirety, we will ultimately come to understand *all things.* This is valuable advice when it comes to myth. Our consumption mindset leads us to believe that *more* is inherently *better,* but there are a great many stories that only reveal their secrets through sustained engagement and slow integration. When we relate with a body of myth in such a way, we allow it to have a deeper and more dynamic impact on our lives. This can have both positive and negative effects – largely dependent upon the merits of the story itself. For this reason, entering the realm of myth is quite *perilous.* The path can lead us toward realization or delusion. Identifying the myths that we allow to inform our lives is a highly sensitive affair.

Myth and Faërie

While myths about gods and heroes are perhaps the best known of our ancestral tales, they often aren't the oldest. Especially in European traditions, some of our most ancient stories are what we now refer to as *fairy-stories* – stories that take place in the enchanted domain of Faërie. We've been conditioned to perceive these as whimsical children's stories and nothing more. But it would be a mistake to think that 'fairy-stories' are simply fantastical 'stories about fairies.' On this, Tolkien writes:

> *It is too narrow... for fairy-stories are not in normal English usage stories about fairies or elves, but stories about Fairy, that is Faërie, the realm or state in which fairies have their being. Faërie contains many things besides elves and fays, and besides dwarfs, witches, trolls, giants, or dragons: it holds the seas, the sun, the moon, the sky; and the earth, and all the things that are in it: tree and bird, water and stone, wine and bread, and ourselves, mortal men, when we are enchanted.*[9]

Fairy-stories may be classified as a kind of myth; they can certainly address 'important' themes, not least *life and death*, but they rarely concern themselves with divine feats and the establishment of institutional authorities. Their perceived frivolity indicates their unsuitability for replacing history or science, and their enchantment is perennially applicable and free from myopic human-centeredness. Much of their magical charm has little to do with magic in its grossest manifestations and more to do with the ennoblement of *common things* by presenting them in an enchanted world. Tolkien writes:

> *... actually fairy-stories deal largely, or (the better ones) mainly, with simple or fundamental things, untouched by Fantasy, but these simplicities are made all the more luminous by their setting. For the story-maker who allows himself to be 'free with' Nature can be her lover not her slave. It was in fairy-stories that I first divined the potency of the words, and the wonder of the things, such as stone, and wood, and iron; tree and grass; house and fire; bread and wine.*[10]

This is an enchanted backdrop against which *all* our stories might feasibly be told. It is perennially applicable to our lived realities, but not *allegorical* or limitingly *topical*. This was a very important point for Tolkien,[11] and it offers an important lesson for us today. *Allegory kills enchantment*. It is only truly useful for aggrandizing one writer's views on one specific set of circumstances. Fairy-stories don't attempt to fictionalize contemporary events, but to expand the boundaries of our perceptible world into the domain of Faërie.

When it comes to approaching our current ecological crisis, especially in art, it's important that we don't simply concoct clever allegories and metaphors for our climate doom. We need stories that are founded upon meaningful relationships between human and non-human persons. As Tolkien notes, '... one of the primal "desires" that lie near the heart of Faërie... [is] the desire of men to hold communion with other living things.'[12]

But this domain of enchanted relationship is, importantly, not a 'supernatural' realm: '*Supernatural* is a dangerous and difficult word in any of its senses, looser or stricter,' Tolkien writes. 'But to fairies it can hardly be applied... for... they are natural, far more natural than [man]. Such is their doom.'[13]

Bardic Revelation

Other stories that are founded upon meaningful relationships between human and non-human persons come in the form of epic poetry. The Tibetan *Epic of King Gesar* is considered by some to be the longest epic poem ever 'composed' and has been a collective project spanning many generations of storytellers, though it remains a *living* tradition. It appears to have its roots in the Tibetan renaissance of the 11th to 13th centuries, but new 'chapters' are being added all the time, with hundreds of

'Gesar bards,' known as *bäp-drung* (Wyl. *'bab sgrung*) – literally '[those unto whom the] myths (*drung*) descend (*bäp*)' – operating in the Tibetan world to this day. It's said that every bard was once a living character in the legend itself, making their mythic accounts in fact *memories* that figuratively *fall into their laps*.

King Gesar is, as a legendary character, believed to have lived around 1,000 years ago, and *may* have once been based on a real historical figure, or more likely a combination of figures, but most of the legend's expanding content comes not from historical accounts but a revelatory process of *mythopoeia*.

Most of us might seek to capitalize on such remarkable narrative creativity and take our stories to the nearest publishing house, but this is rarely the approach taken by bards and other revelatory figures in Tibetan societies. Dr Do Dasel Wangmo, a nun and physician in eastern Tibet, operated as a secret Gesar bard and 'treasure revealer' up until her death just a few years ago. She viewed her special talent as a burden rather than a gift, and wrote down her revelations only to toss them into the fire. Recording the materials was, for her, a matter of necessity – even of life and death. By writing them down, she could exorcise their influence and stop the downloads from relentlessly recurring. She feared that if she were to ignore her bardic destiny entirely, her life-force would be extinguished, but she had no interest in distributing or sharing her work, preferring instead to let it go up in flames so she could focus on her work as a doctor.[14]

Treasure

Revelation and mythopoeia play a central role in Tibetan religious culture. Despite the scant evidence of a historical Padmasambhava, who is presumed to have lived in the 8th century, his *ever-expanding* body of teachings has formed a vital core of Tibetan Buddhist society for over 800 years. While there are traces of a minor figure known as Padmasambhava in texts dating to the 10th century, stories of his role as the quintessential Tibetan Buddhist patriarch only began to emerge in the 12th century, popularized in the works of Nyangral Nyima Özer (*c.*1124–92CE). Özer presented himself as a reincarnation of Padmasambhava's patron, the great King Trisong Detsen, and over the course of his life he composed a rich body of myth surrounding this 'previous life.' Most notably, he writes of Trisong Detsen's (previously undocumented) invitation to Padmasambhava to establish tantric Buddhism in Tibet, elevating him from a fringe figure to the central hero of the Tibetan Buddhist tradition. While these accounts were only 'revealed' around four centuries after the events they portrayed, they had a radical and persistent impact on perceptions of Tibetan culture, creating a durable *charter myth* for a distinctly *Tibetan* form of Buddhism.

This wasn't merely a process of reimagining the past – the Padmasambhava mythos allowed Tibetans to *indigenize* Buddhism and establish a living native *canonical* tradition. This was formalized through a genre of sacred literature known as *terma* (Wyl. *gter ma*), 'treasure,' thought to be hidden by Padmasambhava in the distant past to be uncovered at the most opportune moment in the future. Some *termas* are said to have been physically buried or concealed in the landscape (a landscape in which there are, in fact, many buried artifacts), where they are seen to have been entrusted into the care of powerful local guardians.

While physical *termas* receive the most recognition, most are seen to be 'hidden' in the mind-streams of Padmasambhava's past students, who, not unlike the Gesar bards, are destined to 'retrieve' them at a specific point in a future life.

Miraculous stories of treasure revelation abound in the Tibetan world, and many of the most important *termas* have only been 'discovered' in the past few centuries. Chögyal Namkha'i Norbu Rinpoche (1938–2018), an acclaimed contemporary *tertön* (Wyl. *gter ston*), treasure-revealer, received many of his treasure teachings in dreams. He'd receive 'downloads' of entire liturgies, melodies, and even dances in his sleep, many of which went on to form core practices among his students.

It's common for *tertöns* to report receiving 'lists' of treasures that they're slated to reveal in their career, often imparted by the *dakinis*, female wisdom beings, charged with guarding their secret. In many cases, their revelations are codified in the dakinis' secret 'liminal language,' which only the treasure-revealer can decipher when they encounter it in visions, clouds, the patterns on a cliff face, or any number of symbolic phenomena.

Even if we assume that these treasures are *not* actually downloaded from an 8th-century tantric wizard, the *terma* tradition opens a critical space for religious innovation based on insight and authentic inspiration. It also challenges our perceptions of the boundary between 'revelation' and 'creative inspiration.' It's quite likely that the actual subjective experience of being a revelatory bard is not so dissimilar to the experience of being a visionary artist or inspired storyteller. If one is brought up in a world where memories of past lives and downloads from enlightened beings are perfectly possible, then it would certainly be reasonable to wonder if a stroke of inspiration or genius is actually some kind of revelation. But if the same individual were brought up in a world where revelations were deemed to be impossible, then they might just as reasonably assume their downloads to be a creation of their own imagination. For this reason, notions of 'truth' and 'fiction' may be very complicated, even for a prophet or myth-writer themselves. Some may truly be just as uncertain as anyone else about the veracity of their claims – left to guess for themselves whether their creations are invented falsehoods or revealed truths.

With the commodification of both myth and spirituality, it's hard to know what will come of traditions like *terma* in the future, but over the past millennium it has given us some of the most remarkable philosophical and spiritual works in human history.[15]

Twentieth-Century Mythopoeia

As previously discussed, the field of philology revolutionized the humanities, demonstrating that our cultures are all part of a deeply interwoven tapestry, and that by studying their evolution we can learn more about our history and our connections to one another.

But by the mid-20th century, philology had gained a troubled reputation. Numerous European fascist movements relied on pseudo-philological propaganda to justify their nationalist ambitions.[16] For most philologists, this was a gross perversion of scholarship. The lessons of the 20th century should remind us that even good data can be used by manipulative people for very bad ends. But even with the heyday of philology coming to an end, there emerged the most famous philologist who has ever lived: J.R.R. Tolkien.

As an expert in *comparative* philology, Tolkien was particularly interested in 'filling in the gaps' of historical linguistic knowledge through the process of reconstruction. But for his original works, Tolkien wasn't interested in simply 'recreating' the languages or stories of the past. 'No one would learn anything valid about the "Anglo-Saxons" from any of my lore,' he once wrote. 'My aim has been the basically more modest, and certainly the more laborious one of trying to make something new.'[17]

Rather than focusing on one single cultural system, Tolkien took inspiration from a diverse array of traditions to 'reveal' his *own* body of 'proto'-mythology, viewing his own *subjective* participation as a 'sub-creator' to be an essential ingredient of true mythopoeia. To compose an authentic 'fairy-story,' one needs more than just a knowledge of other fairy-stories – one needs a direct experience of *enchantment*. For Tolkien the philologist, language was the most immediate gateway to this experience – a special form of *evidence* that allowed him to uncover the hidden stories embedded in the spoken (and written) word.

With an understanding of the philological and arguably *mystical* nature of Tolkien's work, his 'fictional' mythology takes on a rather different light. While some of his books saw tremendous success in his own lifetime, fame and fortune were never his intended goals, nor was *The Lord of the Rings* his *magnum opus*. *The Silmarillion*, which remained unpublished until five years after his death, was a far more intimate and ambitious endeavor than the hobbit stories we know and love. It was only due to the tireless efforts of his son Christopher that most of Tolkien's works have now been posthumously published, including long volumes of early drafts and manuscripts annotated with Christopher's 'historical' commentary. This has provided us not only with wonderful stories, but also profound insights into the process of mythopoeia – insights that might help us better understand how 'important stories' come into being, and how myths can both inspire and emerge from the experience of natural enchantment.

The Bard of Middle-Earth

From early on, Tolkien had an ambitious desire to positively impact the world through literature. He was inspired by works like *The Kalevala*, Elias Lönnrot's collection of Karelian and Finnish folklore, which established a powerful sense of Finnish national identity and helped to create a case for their political autonomy. While Britain's circumstances were very different, Tolkien did experiment with the construction of a distinctly *English* mythology, based not on Brythonic Arthurian legends but Anglo-Saxon lore. But over time, the scope of his project expanded significantly, and he abandoned more nationalistic themes for far deeper and global matters. [18]

While he had been inventing languages since childhood, Tolkien's process of mythopoeia began during his undergraduate years at Oxford, when he found himself enchanted by a line of Old English poetry. It was a simple couplet found in *Crist I* (falsely attributed to the poet Cynewulf), which reads:

Éala éarendel engla beorhtast / ofer middangeard monnum sended.
'Hail, Éarendel, brightest of angels, above the middle-earth sent unto men.[19]

Most will recognize *middangeard*, or 'Middle-earth,' as Tolkien's name for his invented world – but, much like the 'world' he was attempting to represent, this word was in no way his 'invention,' but a once-ubiquitous name for our own *inhabited world* in Germanic tongues.

But far stranger and more captivating to the young scholar was the mysterious name *Éarendel*, an old title for Venus appearing as the morning star. Like a treasure-revealer deciphering a mysterious liminal script, Tolkien became enchanted by this line. 'There was something very remote and strange and beautiful behind those words,' he later wrote, 'if I could grasp it, far beyond ancient English.'[20] Éarendel (later spelled *Eärendil*) became the very first character in Tolkien's mythos, making his debut in the 1914 poem 'The Voyage of Earendel the Evening Star.' This marked the 'beginning of Tolkien's own mythology.'[21] After completing his degree at Exeter College, Tolkien grudgingly enlisted in the British Army and joined World War I. He was deployed to France in 1916 and fought in the Battle of the Somme, from which he barely emerged alive. He returned to England with a serious bout of trench fever, and quickly discovered that most of his school friends had died in the battle. It was in this vulnerable state that he returned to his nascent mythology.

His earliest works dealt largely with 'faëries' (later regarded as 'elves'), situating his work in a lineage of European re-enchantment movements that emerged in reaction to industrialization. While he never asserted his stories as factual historical accounts of such beings, he nevertheless regarded them as 'real' agents in our environmental histories – albeit less frequently encountered than they once were when 'a family had fed on the produce of the same few miles of country for six generations.'[22] His stories arose organically out of this fundamentally animistic premise, free from contrivance, as if they were being revealed rather than constructed:

> *[The stories] arose in my mind as 'given' things, and as they came, separately, so too the links grew. An absorbing, though continually interrupted labour... yet always I had the sense of recording what was already 'there', somewhere: not of 'inventing'.*[23]

Just as an experimental archeologist might try to build a dwelling with traditional tools to learn more about its construction, Tolkien placed himself *inside* the process of mythopoeia to experience the development of language and myth first-hand. He had a vivid and active dream life, and it's very likely that lucid dream experiences inspired many of his stories and languages. A particularly prominent motif was, as he describes, '... the dreadful dream of the ineluctable Wave, either coming up out of a quiet sea, or coming in towering over the green islands.'[24] This 'Atlantis complex' (or 'Atlantis haunting') plagued him well into his forties, and served as an important creative basis for his own writings on the fall of Númenor, which he began around 1936.[25] But even at late as 1964 he noted that, 'It still occurs occasionally, though now exorcised by writing about it.'[26]

Perhaps the most illustrative example of Tolkien's 'revelatory' process surrounds the inception of *The Hobbit*. While grading term papers on a summer day in his home in Oxford around 1930, he turned the page of an exam booklet to find an entirely blank page – an unexpected delight, as he considered marking exams to be a 'very laborious' and 'boring' activity. 'I nearly gave an extra mark for it – an extra five marks, actually,' he once joked in an interview.[27] But without thinking twice, he put his pen to the blank page and scribbled the words, 'In a hole in the ground there lived a hobbit.' This, of course, led to his two most famous works, *The Hobbit* and *The Lord of the Rings*, which became two of the main narrative legends set in his Middle-earth.

It's not difficult to see similarities between Tolkien's experience and the experiences of Tibetan revelatory bards and treasure-revealers. The primary difference is simply that Tolkien explicitly characterizes his work as *mythopoeic* and *fictional*, while revelatory bards generally characterize theirs as *historical* and *factual*. While most of us would assume this to be a simple difference between honesty and deception, the reality is far more complex. If you're brought up to perceive past-life memories and revelations as 'real' phenomena that one can potentially experience, then a recurring dream or visionary experience might be surprisingly difficult to classify, even for the experiencer themselves. The reality is that both Tolkien and Do Dasel Wangmo relied on a certain degree of guesswork to decide whether their creative visions were 'true' or 'false.'

According to numerous oral accounts, Chögyam Trungpa Rinpoche, a famous (and deeply controversial) Tibetan lama of the 20th century, once remarked, '*The Lord of the Rings* is a kind of *terma*.'[28] This is a rather extraordinary statement, especially considering that Trungpa identified as a treasure-revealer himself. While he overtly seems to be suggesting, especially to his Buddhist students, that Tolkien accomplished some authentic kind of 'revelation' in his work, the inverse can equally be said. The line between revelation and invention is often phenomenally blurry.

Dangerous Myths

While fantastical stories can guide and heal us, they can also obviously divide us, and there are important pitfalls that we should avoid in their unwitting *application* in our lives. Tolkien writes:

> *Fantasy can, of course, be carried to excess. It can be ill done. It can be put to evil uses. It may even delude the minds out of which it came... Men have conceived not only of elves, but they have imagined gods and worshipped them, even worshipped those most deformed by their authors' own evil. But they have made false gods out of other materials: their notions, their banners, their moneys; even their sciences and their social and economic theories have demanded human sacrifice.*[29]

We should be all too familiar with such false divinities. The *god of capitalism* is, today, the most supreme being in most pantheons. It is *his* call that we heed above all else, calibrating our moral compass to the magnetic pull of profit. He's truly nothing more than a modern fantasy gone terribly awry. He is a 'god,' perhaps, only in the *Gnostic* sense – an evil demiurge who has bound us within his illusory sub-universe, devilishly obscuring our perception of the truth.

But the worship of this god takes place on the grounds of a far more insidious myth. The basic story of *man vs nature* underlies many of our most destructive institutions. While our pride and hubris have led us to believe that a human-centered world will existentially satisfy us, the truth is that it provides nothing but alienation and misery. This isn't an 'authentic myth' that can be used to clarify reality, but a fallacy used to justify our pursuit of excess at the expense of all *others*. Myths such as these are perpetuated through our dissociation from *truth* itself.

In a frightening portent of our modern times, Tolkien predicted:

> *If men were ever in a state in which they did not want to know or could not perceive truth (facts or evidence), then Fantasy would languish until they were cured. If they ever get into that state (it would not seem at all impossible), Fantasy will perish, and become Morbid Delusion.*[30]

Alas, it seems that we are at just such a point in our human story. Our rejection of facts and evidence *on principle* stands directly in the way of our capacity for enchantment and real *connection* with the world around us. Our myths have already morphed into 'alternative facts' and 'morbid delusions,' and before we can effectively write a new story for ourselves, we need to contend with our broken relationship with 'reality.'

Tolkien in the Anthropocene

Tolkien was well aware of the importance of our relationship with the Earth, and a keen critique of industrialization is a major recurring theme in his works. It was only at the end of his life that its true environmental consequences were beginning to come to light, but for him, there was no need to learn that felling trees and burning coal were 'bad' for the future of the planet. He *felt* this intrinsically. 'I am (obviously) much in love with plants and above all trees,' he wrote, 'and always have been; and I find human maltreatment of them as hard to bear as some find ill-treatment of animals.'[31]

This basic animistic ethic can be easily found in his many descriptions of plants and forests. Over 140 types of plants are explicitly described in his work, the vast majority of which are *real* species, but even the fictional ones are largely grounded in legitimate botanical principles.[32] In the last remarks to be published in his lifetime, he wrote:

> *In all my works I take the part of the trees as against all their enemies. Lothlórien is beautiful because there the trees were loved; elsewhere forests are represented as awakening to consciousness of themselves. The Old Forest was hostile to two-legged creatures because of the memory of many injuries. Fangorn Forest was old and beautiful, but at the time of the story tense with hostility because it was threatened by a machine-loving enemy.*[33]

This sensibility toward the welfare of plants is certainly not the central 'moral' of Tolkien's stories, but this makes it even more captivating. Sentient trees and responsive landscapes aren't *anomalies* in Middle-earth, rather an accepted feature of the living world.

Naturally, Tolkien's work was never intended to be an *allegory* for the Anthropocene, particularly because he died long before it was first proposed. But it's phenomenally easy to 'apply' works like *The Lord of the Rings* to our current circumstances. While many races of beings had a role to play in the liberation of Middle-earth, neither the gods, nor the eagles, nor the wizards from beyond the West could effectively vanquish darkness through power and might. The imposition of one's own will upon another – as exemplified by Sauron's creation of the One Ring – is the most essential manifestation of malevolence in this story. The lust for power and dominance over not only people but also forests, rivers, mountains, even life and death is *precisely* the catalyst for what Tolkien presents as 'evil.' In his work, there is a very fine line between seeking to *dominate* the world and seeking to *save* or *heal* it – both depend upon a perception of separateness, and a willingness to self-righteously undermine the agency of others.

The heroes who ultimately manage to oppose this force are neither humans nor elves, but humble and unimposing *hobbits*. With utter disinterest in the pursuit of domination, only the heart of a hobbit could rise to the challenge of destroying the One Ring. In some ways, Tolkien presents us with a *different path*, one in which we retain our *animistic* spirit and shun the pursuit of dominion. *The Lord of the Rings* isn't a story about mighty warriors *overpowering* evil, it's a story about the potency of selfless compassion at what seems like the end of the world. In the bowels of the Anthropocene, stories such as these have a remarkable potential to steer us toward a better future.

Mythic Evolutions

Myths are most effective when they are dynamic and flexible – evolving with their retelling and the expansion of our contextual knowledge. But when we sanctify them and mistake them for infallible 'facts,' we close ourselves off from the potential reception of new and unexpected knowledge.

Part of Tolkien's own efficacy as a myth-writer was his capacity to proactively undermine any such fixity. In the late 1950s, he began revising many of his most cherished and personal myths to ground them more securely in our own world. In his younger years he wrote beautiful 'Elvish' tales about how the Earth was *made round* after once being flat, and how the sun was grown from a sacred terrestrial tree – perfectly respectable *myths*, one might think, but for Tolkien they created too great of a distance between Middle-earth and the planet we know. Rather than framing such stories as the records of gods or immortal elves, he ultimately decided that they must have been 'Mannish myths' – and thus not historical accounts – even within the world of the story itself.[34] Myth and reason needn't be mutually exclusive. In fact, their symbiosis can be incredibly powerful. As Tolkien writes:

> *Fantasy is a natural human activity. It certainly does not destroy or even insult Reason; and it does not either blunt the appetite for, nor obscure the perception of, scientific verity. On the contrary. The keener and the clearer is the reason, the better fantasy will it make.*[35]

Allowing our myths to evolve doesn't need to be a disenchanting process. When guided by logic, reason, and inspiration, such processes may even facilitate breakthroughs in our understanding of reality.

Our attachment to outdated *anthropocentric* myths only perpetuates our blindness to the more-than-human world, thus obscuring our path to recovery. Whether scientific, literary, or religious in nature, the *human-centered story* has well outrun its course. We need plants, animals, and ecosystems to step out of the background and into the fore of our awareness. Even 'nature spirits' like *elves*, *nāgas*, and *kachinas* may have a valuable role to play, reminding us that even in the remotest and deepest parts of the world there is life both seen and unseen. We're *all* sentient beings in mutual relationship, and it's largely through *stories* that such a truth can be psychologically integrated.

Recovery, Escape, and Liberation

Tolkien posited that there were three essential components of any fairy-story: *recovery, escape,* and *consolation.*[36] He describes *recovery* as simultaneously the 'regaining of a clear view' and the 'return and renewal of health,' positing that such tales can help us to recover an experience of the world that is untainted by 'possessiveness,' or the trite 'familiarity' that arises as a consequence of 'appropriation.'[37]

Escape, he argues, is another indispensable feature of the fairy-story, though he's careful to distinguish between the 'Escape of the Prisoner' and the 'Flight of the Deserter.'[38] Escapism needn't always be seen as a bad thing – what is *treatment* if not an escape from disease? He asks, 'Why should a man be scorned, if, finding himself in prison, he tries to get out and go home? Or if, when he cannot do so, he thinks and talks about other topics than jailers and prison-walls? The world outside has not become less real because the prisoner cannot see it.'[39] Stories can indeed offer us glimpses of a world far beyond the limitations of our existing systems, helping us to imagine new ways of living and engaging in the world.

The fairy-story's *consolation* is offered through what Tolkien terms a *eucatastrophe,* the 'good catastrophe,' which provides 'a fleeting glimpse of Joy, Joy beyond the walls of the world, poignant as grief.'[40] This is the *happy ending,* of course, but one that remarkably appears just when it was least expected. In the context of our climate crisis, it may be difficult to trust in such an ending – and if we try too hard to force it, then it may not materialize at all. But for each of us, individually, a sense of *consolation* can be found in the very path of recovery itself. We may not be able to *save the world* – especially if 'the world' just means 'the status quo.' That world was always destined to crumble. But we can allow for new life to emerge from the cracks – new ways of *being* that can help us recover the rapturous joy of being alive. We may not find Elysium beyond our prison walls, but in liberation itself we can find *consolation*.

While the transformative power of a myth or fairy-story can be highly personal, sharing such stories is a powerful means of establishing a sense of connection and collective meaning. Many of the world's greatest tales were once *sung*, often in a group or with participatory elements, to bring people together and establish a sense of communal meaning and identity. To recover from our existential planetary disease, we need to read and tell stories, dabble with fantasy, and seek to illuminate the intersections of the human and more-than-human worlds. We need less corporate myths and conspiracy theories, and more folktales about river-daughters and talking trees.

Those among you who bear the gift and burden of *revealing* and *crafting* stories will have a profoundly sacred and important role in the co-creation of our future. As Val Plumwood writes in 'Nature in the Active Voice':

> *Free up your mind, and make your own contributions to the project of disrupting reductionism and mechanism. Help us re-imagine the world in richer terms that will allow us to find ourselves in dialogue with and limited by other species' needs, other kinds of minds. I'm not going to try to tell you how to do it. There are many ways to do it… The struggle to think differently, to remake our reductionist culture, is a basic survival project in our present context. I hope you will join it.*[41]

We need artists, scientists, scholars, 'influencers,' teachers, and nature-lovers of *all* kinds to work together to recover the stories of the living world and to remind us of our fundamental kinship with non-human sentient beings.

We are arguably living in *the* most important chapter in our entire human history to date, and the weight of responsibility is now incomprehensibly heavy. It is only with hindsight, perhaps in 100 years' time, that we will realize the great influence that the stories we tell *today* will have on our fate as a species. They will either shake us loose from our anthropocentric bondage or drive us deeper into the mire – but at least for now, the choice remains ours.

METHODS OF TREATMENT

The Eightfold Path
of Natural Recovery

As with most diseases, our ecological recovery can be significantly aided by shifts in our behavior – but ours is a *relational* disease, meaning that our treatment cannot be solely inward-focused and we must pay close attention to our behaviors toward *others*. We must also be open-minded and consistent in our approach. As any doctor will tell you, it's not always easy to get patients to do the things that are best for them. You can warn somebody about the dangers of eating too much sugar, but it's often only once their body presents an ultimatum that they genuinely start to *care*.

Val Plumwood argues in *Feminism and the Mastery of Nature* that our relationship with the environment is in fact poorly served by human 'morals.'[1] Rather than being rooted in 'right' and 'wrong,' an effective ethical paradigm *must* be grounded in the cultivation of care. The law isn't what prevents most of us from committing crimes like murder, but rather our basic sense of empathy toward other human beings. Even if fear of punishment *is* the only reason we avoid breaking the law, this still often boils down to a fear of imprisonment and separation from the people we love. We are compelled by *care*, not by *rules*. Thus, casual 'victimless crimes' are quite impossible to eradicate without the very strictest of draconian measures, simply because if we aren't aware of a victim, we won't be compelled to care.

So, while we certainly need laws to safeguard the welfare of nature and non-human beings, this cannot be the *primary* deterrent. If we simply establish *ecocide* as another 'victimless crime,' or an affront against the abstract 'future of humanity,' we'll simply find ways to exploit loopholes in the law. But if we focus more on the cultivation and expression of care, we might just inspire some meaningful change.

In many ways, this approach reflects the Buddha's *Noble Eightfold Path* – his 'prescription' for the existential ailment of suffering. Rather than just tossing out a list of things *not* to do (he did plenty of this later on), his first offering was a foundational overview of the roadmap to recovery. These principles remain highly relevant to our disease at hand. While based on Buddhist concepts, they can be compatible with any religious or philosophical worldview, so long as we're willing to let go of anthropocentrism and all of our 'master identity' complexes.

Any discussion of 'treatment' requires an understanding of what the 'end goal' is supposed to look like. Because we tend to be quite resistant to change, we might imagine that a simple *return to the old status quo* is sufficient. We love the idea of 'sustainability' of course, but this greenwashing incantation is little more than a clever way of saying that we want to have our cake and eat it too. We want to keep doing what we're doing without causing negative repercussions for humans. This is it. If we really ask ourselves what it is that we're trying to 'sustain,' we may realize that the answer is 'exploitation' and 'extraction.' But recovery demands a different end goal. What we must recover is not the old *status quo*, but our dynamic flexibility and capacity to imagine new ways of living. We need to step back *into the stream* and align ourselves with its natural meanderings.

It is also important that we acknowledge that it's unlikely we'll ever manage to 'save the Earth' (whatever that means). It isn't pessimistic to predict that we will fall short of our climate targets and fail to manufacture a 'sustainable' capitalist ecology, or that our demonstrations and calls for revolution are unlikely to lead to a global utopian society anytime soon. But that doesn't mean that the work itself is fruitless. We may not be able to save the entire planet, but we can still have a meaningful impact on the worlds and beings around us. What we *can* do is transform our own relationships – with the Earth, with one another, and with non-human beings. As a wise wizard said:

> ... *it is not our part to master all the tides of the world, but to do what is in us for the succour of those years wherein we are set, uprooting the evil in the fields that we know, so that those who live after may have clean earth to till. What weather they shall have is not ours to rule.*[2]

Recovery is not a matter of *saving* the world, but of transforming our relationship with it. By committing ourselves to such a path, we can both improve the welfare of those around us and reawaken a real sense of the natural wonder of being alive.

So, how do we get there?

Right View: Returning to the Natural State

At the basis of the Buddha's path to liberation is the concept of *view*, or *worldview*. The way we *see* the world has a tremendous impact on how we engage within it. As we know, the view associated with our dominant political, scientific, religious, and economic institutions is that the world revolves around human beings, and that we are fundamentally separate from – and above – 'nature.' From that starting-point, we may decide to *exploit* it or *conserve* it, based solely on our human whims. It's all just a matter of managing resources. But this kind of view, even in the pursuit of 'conservation' and 'sustainability,' is a road to relapse rather than recovery. We must go beyond domination *and* salvation to meaningfully heal from our disease.

The worldview that we must adopt is one in which both human and non-human beings are recognized as manifestations of 'nature.' We have no power to control nature, but what we *can* control is our own behavior, the stories we tell, and the relationships that we form with others. We must realize that we are not separate from nature, and we most certainly are not above it.

Fortunately for us, it's deceptively easy to return to our 'natural state.' All we need to do is drop our egocentric narratives and allow ourselves to *be*. We think of nature as a place to visit or preserve, or maybe 'reunite' with. But we miss the point if we think of it as something *out there*. Nature is *here* – we never left it and we were never separate from it. So there's nowhere to go and nothing to do, apart from dropping the anthropocentric charade and opening our hearts once more to the experiences of others. If we can let go of our utilitarian fantasies about a mechanistic and inert 'natural world' devoid of magic and mystery, we may find that enchantment naturally dawns *as* our road to recovery.

Our world is teeming with magical vitality – all we have to do is allow ourselves to see.

Right Intention: Rewilding as a Path

In the Buddhist world, a great deal of attention is paid to the *intentions* behind our behavior, but some have noted that Buddhism has a bit of an 'intention fetish.' We tend to think that only *malintent* is negative, but in fact *apathy* can be just as dangerous. We need to do far more than avoid ill-will – we need to actively generate empathy and care, and avoid falling into patterns of indifference. When it comes to our treatment of nature, a valuable 'intention' to hold in our hearts is to *rewild* and *regenerate* – to do our part to support a world in which all beings can exist in their own 'natural' state.

There are many ways in which we can rewild our world, beginning in our own backyard. In contrast to *conservation*, which ultimately seeks to 'conserve' resources for prolonged use, rewilding seeks to withdraw needless human interventions to give non-humans a chance to reclaim and reorganize terrain on their own terms. While reforestation is an important part of this process, rewilding isn't all about forests. All types of enviroments have their own virtues, whether they're jungles, grasslands, peatlands, bogs, or even deserts. There is no one-size-fits-all model for healthy terrain.

Even a small city garden can become a vibrant and dynamic 'wild' space, becoming a home and source of food, medicine, and satisfaction for both human and non-human beings. Larger-scale regenerative agriculture can have an even more dramatic impact, guided not by the pursuit of 'pure' humanless nature, but a return to dynamic *interrelatedness*. Nature is not a static stage, but a matrix of living change. But it isn't just our 'environment' that needs to be rewilded. Perhaps even more importantly, *humanity* needs a return to our own natural wildness. We must remember how to be a part of this world.

The active intent to *rewild* and *regenerate* is a powerful antidote to apathy. Even if we fall short of our aims, holding this underlying motivation can change the way that we choose to live. It doesn't mean that we must disengage from nature and cease all existing relationships with non-human beings – that would be both impossible and quite unnatural. We simply need to accept that there is no 'natural order' over which we dominate. We are, all of us – human and non-human, seen and unseen – doing our best in a messy, unpredictable, and remarkable world. We must resist the urge to take control, and instead take responsibility for our own behaviors. This is a useful intention to hold in our hearts.

Right Speech: Using Life-Affirming Language

Our *language* has a particularly direct impact on the way we perceive the world around us, and this is particularly true when it comes to speaking about non-human beings. I have tried to embrace a more life-affirming approach to language in this book. For instance, instead of referring to a singular non-human being as an 'it,' I have opted instead for the personal pronoun 'they.' The use of this non-gendered pronoun has come under fire in recent years, despite having been a part of the English language for centuries, due to its increasing use as a personal pronoun for many trans and gender-nonconforming individuals. It provides a linguistic tool for acknowledging and respecting personhood beyond the social construct of gender binary.

Without minimizing the importance of this evolution in human communities, I think it's important to address the ways in which even simple linguistic constructs like pronouns can deeply impact the ways that we perceive other persons. There is a great deal of baggage tied to 'he,' 'she,' and 'it.' In humans, 'they' helps us to break free from the dualistic paradigm of gender, and might also be useful for disrupting the languaging of anthropocentrism. It's already commonplace to call our pets 'he' or 'she,' but most other animals, plants, and other organisms are habitually identified as 'it.'

Beyond this small token, there are many ways that we can affirm the agency and vitality of other beings through our speech. We shouldn't shy away from speaking about the wants, needs, behaviors, or lived experiences of non-human beings and holobionts like rivers and forests. Talking about them *as if they are persons* might help us to remember that they *are* persons, and compel us to treat them as such. To engage in Right Speech with our environment is, quite simply, to avoid the tendency to objectify non-humans with our language.

Right Action: As Little Harm as Possible

Our current moral problem is not that we find it *justifiable* to exploit non-human beings, but that we find it *inconsequential*. This is the purpose of our 'unseeing.' Our innocence hinges on the assumption that non-human beings aren't *really* alive in any meaningful sense, thus our treatment of them is not an ethical matter. Even our mammalian kin are denied any basic existential rights. We create myths about natural hierarchies to absolve ourselves of responsibility and guilt. But all we are doing is distancing ourselves from the reality of our nature.

Practically speaking, it's inevitable that our lives are sustained through the consumption of other organisms. Unlike our plant cousins, who produce their own nutrients from sunlight, animals simply cannot photosynthesize.[3] Herbivores metabolize their own cells through eating plant bodies, and carnivores metabolize theirs by eating other animals. We've been *omnivores* for much of our own history, beginning as herbivores and gradually adjusting our diet to support ourselves in a range of climatic conditions. Hunter-gatherers had far more varied diets than their agrarian descendents and counterparts, often consuming hundreds of different species (largely plants) on a regular basis. Today we are far more limited, with maize, wheat, and rice accounting for more than half of the calories we receive from plants, and a far higher proportion of animal-derived foods than has ever been 'natural' for our species.

Humans generally don't *need* to eat other animals. We certainly *can*, but there is a long list of things we *can* do that we probably shouldn't. Some believe that food choice is entirely amoral – that the pursuit of nourishment justifies any and all means. But we know intuitively that this isn't true. Most societies deem cannibalism to be immoral, and even the most ardent American carnivore would likely recoil in disgust at the idea of eating their dog or cat. It's not really an issue of 'diet' being magically exempt from ethical dimensionality, but an issue of how we ascribe value to non-human beings.

Of course, embracing plant sentience ushers in a new range of ethical questions. Should we stop eating plants as well? What about our even closer fungal kin? Should we all aspire to become *breatharians* or simply starve ourselves out of fear that we might cause suffering? It's important that we attempt to answer these questions earnestly, rather than just roll our eyes. As Matthew Hall states:

> *If we acknowledge the sentience of the plant kingdom... and we accept the violence that we take part in, we can be prompted to seek ways to reduce it. Holding both sentience and violence in mind whenever we approach a plant to use its leaves, roots, or shoots, we can restrain ourselves, taking only what we need and not killing wantonly. If we hold this violence close when preparing and eating our food, we may be prompted to eat only what we need and to minimize waste wherever possible.*[4]

In India, Jain philosophers in the first millennium BCE proposed that only the expendable aerial parts of plants, like leaves and fruits, should be eaten, while roots and more vulnerable parts of a plant should be left untouched. This led to strict dietary restrictions for Jain monastics, who avoid eating any foods that are obtained through the death of a plant or animal. But laypeople were simply advised to be mindful of the suffering caused in the procurement of their food, and to seek to cause *as little harm as possible*. This principle can be tremendously powerful, so long as we understand and take responsibility for the parameters of 'harm' itself.

We know that plants' relationship with corporeality is quite unlike that of animals. Plants and fungi don't need to rely on sexual reproduction to produce offspring – most can functionally clone themselves by simply dividing their physical bodies into multiple beings, a process that we commonly assist through the cultivation of cuttings. Whatever sense of 'self' a plant might possess, they appear to eschew centralization in lieu of a more diffused and infinitely divisible awareness. Relationship is not as linear as 'parent' and 'child' – it is more fluid and complex. While many plants (like trees) will usually die if they are chopped down, many others are perfectly accustomed to being periodically mowed down and eaten. This doesn't mean that we should be greedy or entitled when 'harvesting' them for food, but this form of exchange is often a very natural part of what it means to be a plant.

Furthermore, we should bear in mind that most root vegetables are harvested at the end of a plant's natural life cycle, and up until their uprooting they are often allowed to live a life that is more or less consistent with their natural instincts. We *can* effectively provide plants with a natural and comfortable lifestyle while still relying on them for food, which is often a stark contrast to the dynamics of animal agriculture. Unlike a carrot or potato, a captive pig is not allowed to live a normal, natural life prior to their slaughter. And because livestock animals also require food, eating meat ultimately leads to far greater consumption of both plant and animal-based food (as well as other resources like water) in the long run.

In contrast to the Jain approach of *causing as little harm as possible*, the Buddhist tradition unfortunately relied on a plant-blind moral paradigm to ignore the suffering of plants. This approach is based solely on convenience, and such a mentality is particularly problematic when vegetal environments are so profoundly vulnerable. Instead of pretending that plants are inert to make our behaviors more palatable, it would be far more honest for us to grapple with the inevitability of violence as a function of being human. As Hall writes:

Human beings are totally dependent on the plant kingdom for food, energy, and shelter. Our lives are such that we do not have the choice of eschewing the use of plants. The severe dietary restrictions of the Jaina are not a path that many on earth can take. The only way to completely 'avoid' violence toward plants is to go on pretending that they are not sentient, aware, and intelligent....[5]

Animistic hunter-gatherers certainly killed both animals and plants for food, and they continue to do so today. But this killing occurs side by side with the basic acknowledgment of the genuine vitality of the being that is killed. It is on *this* basis that matters of 'necessity' can be morally and spiritually negotiated.

This is remarkably different from the anthropocentric approach, in which the vitality of the non-human is dismissed altogether. 'Livestock' have been subjected to physical and psychological abuse for countless generations, with animal husbandry becoming increasingly unnatural and mechanistic in the rise of industrial agriculture. 'Factory farming' produces a truly incomprehensible amount of suffering.

I'm not saying that killing an animal is morally neutral as long as we acknowledge they're alive, but an appreciation of their conscious experience quite naturally compels us to be less wasteful and cruel in our pursuit of nourishment, which itself could have a dramatic impact on our relationship with non-human beings.

It's interesting to consider paradigms like 'animal sacrifice' in this light. Most westerners would argue that animal sacrifice is barbaric, even if the sacrificed animal is ultimately fed to humans. The sanctification of the process makes us very uncomfortable, perhaps bringing us *too* close to the recognition of an animal's existential significance. A sacrifice is, after all, a *loss* of something, and even if the animal is ultimately eaten, there is an implicit acknowledgment that something has been lost in the process. This makes us uneasy, and as a result, most of us shudder at the thought of a goat's throat being slit at the altar of *Dakshinkāli* – even if we have no qualms about eating a goat whose throat was slit on a factory farm.

Animal ethics are highly complex and go far beyond the question of whether or not we 'should' eat meat. The hunting practices of the Amazonian Achuar or the Ojibwe of Ontario are starkly different from our dominant forms of agriculture. In indigenous contexts, moderation and care are organically motivated through more-than-human social bonds, and a recognition that prey animals are 'persons' in and of themselves.[6] Rather than imposing a one-size-fits-all dietary paradigm, we instead need to become more critical of modern agricultural practices, even and especially considering the prevalence of starvation.[7] Half the world's habitable land is used for agriculture,[8] and of that, a whopping 80 per cent is used for feeding and rearing livestock – *not* for growing human food.[9] But despite how phenomenally resource-intensive animal agriculture is, these animal-based foods only contribute 20 per cent of the world's calories, and only 37 per cent of our protein.[10] The environmental toll of this system is astoundingly disproportionate, decimating entire ecosystems as a mere *side-effect* of treating animals as a commodity. This is fundamentally unsustainable.

In a best-case scenario, we would all naturally transition to local biodynamic agriculture, the sustainable use of wild foods, and responsible fair trade. We would eliminate industrial animal agriculture entirely, and our diets would mostly include a diverse array of ethically harvested plants and fungi. Access to healthy food would be a right, not a privilege, and no human would be forced to starve as a result of the unjust distribution of resources.

In the meantime, we don't need to be perfect to make a difference, and while we certainly need radical reforms in the energy sector, our use of plastics, and many of our patterns of consumption, paying attention to our diet is one of the easiest and most direct ways that we can alter our relationship with non-human beings.

Right Livelihood: 'Sustainability' and Anti-Exploitation

The topic of 'right livelihood' is complex. In the modern world, the capitalist model is one of the most serious co-morbidities accompanying our ecological disease. When combined with non-human disregard and a 'class' paradigm with mythic roots stretching all the way back to Yemo's dismemberment, it should come as no surprise that many view the exploitation of the environment and the working-class as the 'natural order of things.' Moreover, in a capitalist society, the ways that we engage with private industry are immensely consequential – even more so, one might argue, than our engagement with the democratic process. We 'vote' with our dollar and with our labor every day, and it's nearly unavoidable that our time and money will ultimately become entangled with paradigms of natural exploitation.

The regulation of industry needs to be high on our list of methods of averting climate change, but these regulations need to address far more than just human welfare. When we pollute a river with toxic chemicals, we are doing more than compromising human access to clean water – we are committing an act of violence against a massive community of beings. Even uprooting a tree without very good reason should probably have far harsher consequences than 'vandalizing' a building with graffiti, but you're far more likely to go to jail for painting a wall than for killing a tree. If you can *capitalize* on your destruction of nature, then you may even be rewarded with riches, acclaim, and vast political power. In the moral universe of capitalism, *exploitation* is a skill and a virtue.

Consequently, our societies are still rife with *human* exploitation and systemic oppression, and it's not difficult to see the many ways in which our anthropocentric bias has devolved into progressively more restrictive paradigms of oppression and exploitation. Our societal shadows of sexism, racism, classism, xenophobia, and countless other biases remain an ever-present threat to our stability and welfare, and are deeply intertwined with our systems of ecological exploitation. This is abundantly evidenced by the countless indigenous peoples displaced due to climate change, the disproportionate amounts of pollution in areas principally inhabited by people of color, and the widespread inaccessibility to nutritious and organically farmed foods in low-income communities.

Our duty is to dismantle these systems as best we can, and to contribute to fairer and more equitable alternatives. Removing the philosophical justification for them is an important start, but once the foundation is gone, we need to figure out what to do with the pile of rubble that remains. Our therapy quite simply won't work if our very means of sustaining ourselves is perpetuating the problem. Infinite growth is untenable within a finite system, and capitalism is bound to lead to destruction.

By supporting worker-led cooperatives, small-scale biodynamic farms, social services, and other projects aimed at public welfare, we can help little sprouts of life to break through the cracks of late-stage capitalism.

Right Diligence: The Durability of the Heartfelt Ethic

In Buddhism, *diligence* (or *effort*) is described as a kind of motivation that compels us to *do the right thing* and avoid wrongdoing, even in moments of apathy. It is 'putting in the effort,' even when inconvenient, to avoid causing harm. But we cannot be sufficiently motivated by willpower or grit alone – we need to authentically *care*. It is only genuine care that can provide us with the durable motivation to truly invest in the welfare of others.

There's a degree of patience and resilience, as well as forbearance, in this kind of work. We need to actively nurture our relationships, and to keep them warm even if we're feeling rather chilly. Right diligence must, then, be a continual re-investment in our genuine concern for the welfare of others. We need to continually remind ourselves of what's really at stake – not just for *ourselves*, but for everyone around us. We must learn to foster a heartfelt ethic of care, rather than simply subscribe ourselves to some nice-sounding moral paradigm to try and be a good person. We need to generate authentic compassion for other beings.

One of the most impactful instructions I ever received on the cultivation of compassion came from one of my *Chöd* teachers, Drubpön Lama Karma, during a series of teachings in Colorado. While commenting on the differences he'd noticed between

Bhutanese and American culture, he mentioned that our treatment of dogs caught him somewhat by surprise. He was struck by the deep familial *love* that we seemed to feel for our dogs, as well as our rather uncharacteristic patience and forbearance when they misbehaved. 'You should really try to see *everyone* in this way,' he said. 'If we could see *all* beings the way that you see dogs, we would live in a much more compassionate world.'[11]

Right Mindfulness: Paying Attention to Nature's Vitality

In the path to liberation laid out by the Buddha, *mindfulness* is famously one of the most important factors. On a most basic level, mindfulness is the simple practice of becoming attentive to the present moment. By quieting our mental chatter and habitually returning our awareness to a stable object, like our breath, we can learn to become much less distracted and refine our capacity for deeper insight.

In a traditional context, mindfulness is only introduced to students following a foundational training in ethics. Simply becoming *mindful* of things isn't of much use if our moral compass is fundamentally askew. In a modern secular context, this preliminary step is unfortunately often omitted. But it is only with a firm ethical foundation that mindfulness can truly benefit us on an interpersonal level. Corporate mindfulness trainings might temporarily help overworked and underpaid workers relax and feel less stressed, but silencing our stress response won't help us to address an immoral and exploitative system.

When we *start* from a place of genuine care, however, the practice of mindfulness can become a means to remarkable ends, including in our relations with others. As Dan Nixon writes, 'If we're willing to slow down, to pause, to touch in with the pulsing, breathing lifeworld of the body throughout the day, then we can restore our presence, our aliveness, our precious connection with other beings and things no less vital for the fact that they are other to us.'[12]

Mindfulness is not only useful in moments of quiet contemplation. In the Tibetan tradition, most 'meditation' practices are distinctly active, with visualizations, recitations, and ritual activities helping the practitioner engender a state of contemplation *within* dynamic motion.

Perhaps more importantly, ritual also helps us to become mindful of our broader place in a spiritual and ecological cosmos, as one of countless characters in a more-than-human story. As Joseph Campbell once said:

> *A ritual is the enactment of a myth. And, by participating in the ritual, you are participating in the myth. And since myth is a projection of the depth wisdom of the psyche... you are being, as it were, put in accord with that wisdom, which is the wisdom that is inherent within you anyhow. Your consciousness is being reminded of the wisdom of your own life.*[13]

Through ritual engagement we become 'mindful' of not just the thoughts in our head, but also the lived experiences of others.

One of our most ancient rituals, which can also be one of our most valuable, is the simple practice of *offering*. Offering is an expression of generosity, but also an expression of *kinship*, dissolving the boundaries between self and other in acknowledgment of the mutuality of our being-ness. Offering is not strictly transactional

– while there may be some hope for reciprocity, it is usually both amorphous and unmandated. We are simply investing in our collective welfare from a place of generosity.

We don't need ancient rituals to connect with others in this way. Watering your garden can be a ceremonial offering rite, or you could ritually offer some peas to the ducks at a local park. Whenever I visit 'enchanted' or sacred places, or head out to forage for herbs, I always take a bit of oat milk or a bit of homemade incense to offer to the unseen beings. While I once used to recite formal Tibetan prayers with my offerings, I've learned to become more attentive and spontaneous, focusing on establishing *connection* rather than adhering to 'structure' and 'sanctity.' In doing this, I've found that my experience of such environments is far more intimate and personal, allowing for a moment of genuine *mindfulness* to the non-human lives around me.

Every walk in the park is an invitation to become attentive to the more-than-human world. From the weeds creeping out from the pavement to the insects and birds flying overhead, the sheer number of *lives* all happening at once can be truly awe-inspiring. Allow yourself to *take in* the fullness of the living world. Let it fill all your senses, without trying to judge or interpret it in human terms. If we can become mindful of the many *minds* with which we share this planet, we might come to a deeper understanding of the nature of 'mind' itself.

Right Concentration: Imagining a New Future

Myths, fairy-stories, and folktales are, as we have seen, valuable tools for expanding our worldview. When well crafted, they can inspire a healthy sense of natural enchantment and curiosity about our world, our history, and the non-human beings that surround us. But experiences of enchantment are as varied as the minds in which they arise, and each of us is bound to awaken to enchantment in our own unique way. If you're lucky, you may have found a body of *authentic myth* that speaks to your heart. Maybe you've found many, or are still searching, or perhaps you are birthing your own. We think that we need more scientists and data analysts to make better sense of the world, but the truth is we need more artists, storytellers, and mythmakers. To contribute to a brighter future on this planet, we need to learn to *imagine* new and alternative ways of being.

We would be wise to take our cues from the many animistic wisdom-holders who remain with us – not by appropriating and commodifying their traditions, but by taking their knowledge and advice seriously. I have been fortunate to spend much of my life surrounded by holders of such wisdom, and consider myself immensely privileged to have been able to engage meaningfully with multiple systems of traditional knowledge. But the spread of knowledge isn't all about assimilation and appropriation – it's about the creative process of interaction. Knowledge grows and transforms when it is shared. The transmission of knowledge can be an alchemical process, particularly when it occurs across time, space, and cultural environments. This has never been a strictly empirical or logical process, but also a deeply creative one.

'Artists' are, perhaps, like the neurons and 'green matter' that help animals and plants translate data into action. We need artists to translate what we *know* about the world into what we *feel* and what we *do* about it. This doesn't always have to take the form of lofty myths – even pop music can create subtle shifts in our perception of the world. Shortly after I began writing this book, New Zealand artist Lorde (Ella Yelich-O'Connor) released her third studio album, *Solar Power*, describing it as 'an album about the spiritual power of the natural world'[14] inspired in part by her dog, Pearl, whose unexpected death in 2019 delayed the project's completion by a couple of years.[15] A few weeks after the album's release, she released an EP of five of the album's songs interpreted in Te Reo Māori, produced in collaboration with indigenous artists and poets with numerous allusions to Māori lore and knowledge.[16] Overall, the project received mixed reviews from critics: Some found *Solar Power* to be a cliché and overly privileged left-turn for the 'pop' artist – even a 'collection of heat haze hippy noodlings'[17] – while others deemed it a 'near masterpiece'[18] that 'grows in quiet stature with every listen.'[19]

The timing of this album was quite poignant for me. When I began writing this book, I was very worried that it would be too obscure and unrelatable to be taken seriously. But a simple pop album from an artist that I deeply admire gave me a surprisingly powerful boost of courage. Music is powerful – it taps into a deep and even *mythic* place in our psyche and can inspire emboldenment, reflection, nostalgia, even feelings of reverence and awe. It can transport us to happy places, orchestrate revolutions, or simply remind us that we're not alone in our greatest moments of darkness. It's a potent tool, and when it flows from a place of natural enchantment, incredible things can happen.

We *all* need to turn our attention and care toward the more-than-human world in a meaningful way, and we need to do so from many different angles if we want to create authentic change on a systemic level. Whether you're a therapist, a banker, a bartender, a songwriter, or a landscaper, each of us can do our part to bring non-humans out of the background and to the forefront of our awareness. But our recovery isn't actually about 'saving the world,' nor is it about becoming impervious to our environment. It's about transforming our own experience of the world to inspire more authentic and ethical engagement. If we can bring ourselves to truly *see* the *unseen*, we will find that the world has always been far richer and more magnificent than we could ever imagine.

Conclusion

LEARNING TO SEE

Over the course of this book, we've seen how the dangerous myth of anthropocentrism has played a critical role in our growing disregard for nature. We've traced its development through philosophical, religious, and secular scientific movements and seen how it became mischaracterized as a 'rational' worldview for a disenchanted world. While factors like industrialism and capitalism are dangerous co-morbidities that have worsened our condition over time, the gravest ecological consequences we face are primarily rooted in our anthropocentrism.

We've convinced ourselves that humanity is a beacon of reason in a dim and unconscious universe, and that our divine intellect places us in a central existential category of our very own. We may live 'in' nature, but we don't perceive ourselves to be governed by organic laws and limitations. Even as we face mass extinction, ecological collapse, rising temperatures, and the threat of nuclear cataclysm, we remain stubbornly convinced that we are on a permanently upward trajectory. Even on a strictly *evolutionary* basis this is obviously false, since the 'success' of our species is dependent on our ability *not* to destroy our environment. But surely 'survival' and 'happiness' are mere trifles that can be set aside in the pursuit of dominion.

Like an unmanaged addiction, the first step in our recovery is to acknowledge that we have a problem. From this point of sober reckoning, we can then begin to investigate and uproot its underlying causes in an honest way, and ultimately to make amends. An earnest process of self-reflection reveals that it's not just our systems and institutions that make us sick, but the very lens through which we view the world. Our alienation from nature has made us less healthy, less safe, less compassionate, and less fulfilled.

It's hard to be optimistic when looking at the grim state of the world. If the past decade has taught us anything, it's that the divide between 'what we know' and 'what we believe' is perhaps wider than ever before. Many of our myths have already been corrupted into morbid delusions, and our relationship with 'truth' is becoming increasingly tenuous. But we must remember that crisis, ecological or otherwise, is not actually our natural state but a painful symptom of systemic imbalance. The therapeutic process isn't about *adding* new stuff to fill up our lives, but about sloughing off the layers of our diseased and broken worldviews to return to a more authentic mode of being. We may lose our sense of human chauvinism and entitlement in the process, but we will *gain* relationship, enchantment, and a green Earth for generations to come – a reasonable trade-off, if you ask me. As historian Jonathan Coope writes:

> ... *once we begin to expand our conceptions of 'care' toward convivial fellowship with the other-than-human beings with whom we share our destiny on the earth, maybe we will make a happy discovery that,*

in so doing, we also expand the compass of our lived experience... of
wealth, well-being and what it might mean to be fully human.[1]

But dismantling anthropocentrism, and the many systems that have been built around it, will not be an easy task. We've become so habituated to its presence that we mistake it for a basic truth. But if religion is the opiate of the people, then anthropocentrism is ideological methamphetamine, compelling us ruthlessly toward violence, self-aggrandizement, and complete desensitization to the needs and experiences of others. We harm *everyone* through our abuse of this philosophical drug, and yet we don't *want* to change, perhaps because we think that our 'success' as a species is dependent upon our unflinching exploitation of nature.

But there *is* another way. A narrow escape from ecological cataclysm is still possible if we can simply bring ourselves to *see* the inherent value of non-human beings. Luckily for us, changing is what we do best. We are nothing if not a mutable species, inventing and reinventing ourselves time and time again, and often transforming our environment in the process. If we can commandeer this process and imbue it with compassion and enchantment, we might still be able to direct ourselves toward a brighter future. Such an 'animistic turn' for society could inspire more earnest concern for the welfare of non-human beings. This cannot simply be forced upon us, but the promise of authentic connection can inspire and enchant us into a dynamic paradigm shift.

In a perfect world, any notion of 'morality' would be replaceable with knowledge and compassion. The best way to compel people not to cause suffering is simply to get them to *care*. We generally don't need to be *told* not to hurt those we love. So we need a paradigm of bioethics built around empathic connection, not mere data points. Fortunately for us, we are excellent at inspiring empathy – but for this we need artists, storytellers, historians, and philosophers, not only scientists.

We are living beings in a living world, and if we can allow that simple realization to truly *enchant* us, then we can use that experience to move toward ecological recovery. The cultivation of an animistic worldview is invaluable in such a venture, bringing us into direct communion with the many 'intelligences' of the world – those embodied in the plants, animals, forests, streams, oceans, and even the living planet itself. Animism is our most *basic mode of existing* in a more-than-human universe. It's not primitive but *foundational*, and our departure from it was a herald of our demise, not our ascent to greatness. But when we cast off the veils of our own self-obsession, we're bound to realize that we've never actually departed from *nature* itself. It's always been right here – in our cities, our homes, our bodies, and in the luminous awareness of our psyche.

Animism certainly isn't a *panacea* for *all* of humanity's woes, nor should we think that animistic peoples have always lived in perfect egalitarian utopias. Human life is complex and messy, but it's nothing if not mutable. In their recent book, *The Dawn of Everything*, historians David Wengrow and (the late) David Graeber ask:

> *What if, instead of telling a story about how our species fell from some idyllic state of equality, we ask how we came to be trapped in such tight conceptual shackles that we can no longer even imagine the possibility of reinventing ourselves?*[2]

This is really a question worth pondering. What if we begin to tell a different story, one in which we have *never* been separate from the wonders of nature, though some of us lost sight of it for a while. Our world is full of countless strange and wonderful *persons*, and only very few of them are human beings. Life is richer, more fulfilling, and far more enchanting in a more-than-human world – we just need to open our eyes.

NOTES AND REFERENCES

Introduction: The Call of the Unseen

1. Pratchett, T., *Small Gods*, Gollancz, London, 1992, p.6. *See also* Harvey, G., *Animism: Respecting the living world*, 2nd ed., Columbia University Press, New York, 2017, p.23

2. United Nations 2021, 'The Global Forest Goals Report 2021'; https://www.un.org/esa/forests/wp-content/uploads/2021/04/Global-Forest-Goals-Report-2021.pdf

3. Tolkien, J.R.R., *The Silmarillion*, ed. Christopher Tolkien, Allen & Unwin, London, 1977, pp.26–8

Chapter 1: The Hard Truth

1. Pliny the Elder, *Natural History: A selection*, trans. J.F. Healy, Penguin, London, 2004, p.286

2. This notion of 'Peak Humanity' is based on lectures delivered by Dr Martin Savransky at Goldsmiths College, University of London, October 2022.

3. Almond, R.E.A., Grooten, M., and Petersen, T. (eds), *Living Planet Report 2020: Bending the curve of biodiversity loss*, World Wildlife Fund (WWF), Gland, Switzerland, 2020; https://www.zsl.org/sites/default/files/LPR%202020%20Full%20report.pdf

4. Hawken, P., et al., *Drawdown: The most comprehensive plan ever proposed to reverse global warming*, Penguin, London, 2017, p.39

5. Woodhall, A., 'Addressing Anthropocentrism in Nonhuman Ethics: Evolution, morality, and nonhuman beings', doctoral dissertation, University of Birmingham Dept. of Philosophy, 2016, p.23

6. Anthis, K.A., and Anthis, J.R., 'Global Farmed & Factory Farmed Animals Estimates', Sentience Institute, 2019; https://www.sentienceinstitute.org/global-animal-farming-estimates [accessed May 6, 2022]. For more on the environmental toll of industrial fishing and poaching practices, *see* Hill, J., 'Environmental consequences of fishing practices', n.d; https://www.environmentalscience.org/environmental-consequences-fishing-practices [accessed March 20, 2022]; *also* Sebo, J., 'Against human exceptionalism', *Aeon*, May 5, 2022; https://aeon.co/essays/human-exceptionalism-is-a-danger-to-all-human-and-nonhuman [accessed May 6, 2022].

7. Adam, D., '15 million people have died in the pandemic, WHO says', *Nature* 605, (2022): 206; https://doi.org/10.1038/d41586-022-01245-6

8. Schneiderman, J., 'The Anthropocene Controversy' in *Anthropocene Feminism*, ed. R. Grusin, University of Minnesota Press, Minneapolis, MN, 2017, p.170

9. Latour, B., 'How Not to (De-)animate Nature' in *Facing Gaia: Eight lectures on the new climatic regime*, Polity Press, Cambridge, 2017, pp.42–74

10. Bonneuil, C., and Fressoz, J-B., 'Who is the Anthropos?' in *The Shock of the Anthropocene: The Earth, history, and us*, Verso, London, 2015, pp.55–6

11. Woodhall, *op. cit.*, p.174

12. Cited *ibid.*, p.176

13. Solnit, R., 'Big oil coined "carbon footprints" to blame us for their greed. Keep them on the hook', *Guardian*, 23 August 2021; https://www.theguardian.com/commentisfree/2021/aug/23/big-oil-coined-carbon-footprints-to-blame-us-for-their-greed-keep-them-on-the-hook

14. MIT Climate, 'What is the ideal level of carbon dioxide in the atmosphere for human life?', MIT Climate Portal, 2021; https://climate.mit.edu/ask-mit/what-ideal-level-carbon-dioxide-atmosphere-human-life

15. UC San Diego publishes data from Mauna Loa Observatory, presented in a number of useful charts, on their Keeling Curve webpage: *see* https://keelingcurve.ucsd.edu.

16. IPCC, 'Global Warming of 1.5°C: An IPCC Special Report on the impacts of global warming of 1.5°C above pre-industrial levels and related global greenhouse gas emission pathways, in the context of strengthening the global response to the threat of climate change, sustainable development, and efforts to eradicate poverty', V. Masson-Delmotte, P. Zhai, H-O. Pörtner, *et al.* (eds), Cambridge University Press, Cambridge, 2018, pp.541–62; doi:10.1017/9781009157940.008

17. Klein, N., *This Changes Everything: Capitalism vs. the climate*, Simon & Schuster, New York, 2014

18. Ibid.

19. Tolkien, J.R.R., 'On Fairy-Stories' in *The Monsters and the Critics and Other Essays*, HarperCollins, London, 2006, p.117

20. Josephson-Storm, J., *The Myth of Disenchantment: Magic, modernity, and the birth of the human sciences*, University of Chicago Press, Chicago, IL, 2017

21. Hall, J., *Ernest Gellner: An intellectual biography*, Verso, London/New York, 2011

Chapter 2: The Natural State

1. Shakespeare, W., *The Tempest*, Act IV, scene i, in *The Annotated Shakespeare*, eds B. Raffel and H. Bloom, Yale University Press, New Haven and London, 2006, p.107

2. Woodhall, A., 'Addressing Anthropocentrism in Nonhuman Ethics: Evolution, morality, and nonhuman beings', doctoral dissertation, University of Birmingham Dept. of Philosophy, 2016, p.178

3. Sagan, C., Margulis, L., and Sagan, D., 'Life', *Encyclopaedia Britannica*; https://www.britannica.com/science/life [accessed June 26, 2022]

4. Sahlins, M., 'What kinship is (part one)', *Journal of the Royal Anthropological Institute* 17, (2011): 2–19. *See also* Hall, M., *The Imagination of Plants: A book of botanical mythology*, SUNY Press, Albany, NY, 2019, p.xxvii

5. Boussard, A., Fessel, A., Oettmeier, C., *et al.*, 'Adaptive behaviour and learning in slime moulds: the role of oscillations', *Philosophical Transactions of the Royal Society B* 376, (2021): 18–20; doi: https://doi.org/10.1098/rstb.2019.0757

6. Marder, M., 'The life of plants and the limits of empathy', *Dialogue* 51, (2012): 259–73. *See also* Hall, M., *The Imagination of Plants: A book of botanical mythology*, SUNY Press, Albany, NY, 2019, p.xxvi.

7. Hall, M., *ibid.*, p.xxvii

8. McKie, R., 'Loss of EU funding clips wings of vital crow study in Cambridge', *Guardian*, 28 May 2022; https://amp.theguardian.com/environment/2022/may/28/loss-eu-funding-crow-study-cambridge-brexit-corvid?fbclid=IwAR14BshF9Ldewb344EXTJ3Z Mbqj-kStnQlkoCi6CMNy5YZAyLWKtncjdCU0 [accessed May 20, 2022]

9. Watanabe, S., Sakamoto, J., and Wakita, M., 'Pigeons' discrimination of paintings by Monet and Picasso', *Journal of the Experimental Analysis of Behavior* 63 (2), (1995): 165–74; doi: https://doi.org/10.1901/jeab.1995.63-165

10. Mancuso, S., *The Revolutionary Genius of Plants: A new understanding of plant intelligence and behavior*, Simon & Schuster, New York, 2017, pp.87–9

11. There are numerous approaches to determining the date of domestication, but a 2017 genetic analysis supports a primary domestication event occurring sometime between 20,000 and 40,000 years ago. *See* Botigué, L.R., Song, S., Scheu, A., *et al.*, 'Ancient European dog genomes reveal continuity since the Early Neolithic', *Nature Communications* 8, (2017): 16082

12. There is no reason to think of this as a purely human-driven development, and thus allegations that we 'domesticated' dogs are actually quite problematic. Co-habitation between humans and wolves was a mutual process that conferred significant benefits on both parties.

13. *See* Marshall-Pescini, S., Schaebs, F.S., Gaugg, A., *et al.*, 'The role of oxytocin in the dog–owner relationship', *Animals (Basel)* 9, (2019): 10; doi: 10.3390/ani9100792. The fact that dogs, as well as humans, have evolved to release oxytocin during close human–dog interactions gives us every reason to assume that they perceive our relationships to be a source of emotional pleasure and satisfaction, and that this is deeply biologically engrained.

14. A study of Chaser, a border collie, demonstrated that she had successfully learned over 1,000 words: Pilley, J.W., 'Border collie comprehends sentences containing a prepositional object, verb, and direct object', *Learning and Motivation* 44 (4), (2013): 229–40; https://doi.org/10.1016/j.lmot.2013.02.003. Multiple studies have observed a canine capacity for 'fast mapping' and inferential reasoning, and there is also a growing movement of using 'buttons' programmed with pre-recorded words to train dogs to 'speak' to their humans, a trend that has become highly popular on social media platforms like TikTok.

15. Taçon, P., and Pardoe, C., 'Dogs make us human', *Nature Australia*, Autumn 2002

16. Janssens, L., Giemsch, L., Schmitz, R., *et al.*, 'A new look at an old dog: Bonn-Oberkassel reconsidered', *Journal of Archeological Science* 92, (2018): 126–38

17. Losey, R.J., Bazaliiskii, V.I., *et al.*, 'Canids as persons: Early Neolithic dog and wolf burials, Cis-Baikal, Siberia', *Journal of Anthropological Archeology* 30, (2011): 174–89

18. Siniscalchi, M., d'Ingeo, S., Quaranta, A., 'Orienting asymmetries and physiological reactivity in dogs' response to human emotional faces', *Learning & Behavior*, 46, (2018): 574–85

19. Mallory, J.P., and Adams, D.Q., *The Oxford Introduction to Proto-Indo-European*, Oxford University Press, Oxford, 2006, p.439

20. Brenner, E.D., Stahlberg, R., Mancuso, S., *et al.*, 'Plant neurobiology: an integrated view of plant signaling', *TRENDS in Plant Science* 11 (8), (2006): 413–19

21. *Ibid.*, 413

22. Gagliano, M., *Thus Spoke the Plant: A remarkable journey of groundbreaking scientific discoveries and personal encounters with plants*, North Atlantic Books, Berkeley, CA, 2018, pp.33–4

23. Hall, M., *Plants as Persons: A botanical philosophy*, SUNY Press, Albany, NY, 2011, p.148

24. Ibid.

25. Baluška, F., Mancuso, S., and Volkmann, D., *Communication in Plants*, Springer-Verlag, Berlin, 2006, p.28. *See also* Hall, M., *ibid.*, 2011, p.147.

26. Hall, *ibid.*, p.148

27. Mancuso, S., *The Revolutionary Genius of Plants*, Simon & Schuster, New York, 2017, p.xi

28. *Ibid.*, pp.8–10

29. Gagliano, *op. cit.*, pp.55–71

30. Mancuso, *op. cit.*, pp.5–15

31. Calvo, P., *Planta Sapiens: Unmasking plant intelligence*, The Bridge Street Press, London, 2022, pp.7–12

32. *Ibid.*, p.12

33. *Ibid.*, pp.13–14

34. Mancuso, *op. cit.*, pp.44–5

35. Ibid.

36. Ibid.

37. Wohlleben, P., *The Hidden Life of Trees: What they feel, how they communicate – discoveries from a secret world*, trans. Jane Billinghurst, Greystone Books, Vancouver/Berkeley, CA, 2016

38. It should be noted that there are certain smells that humans are particularly adept at identifying, even in very low concentrations, like the smell of freshly fallen rain on dry earth (also known as *petrichor*). Our sensitivity to this meteorological signal is more acute than a shark's sensitivity to the smell of blood.

39. Harrod Buhner, S., *Plant Intelligence and the Imaginal Realm: Into the dreaming of Earth*, Bear & Company, Rochester, VT, 2014, p.127. *See also* Choh, Y., Kugimiya, S., and Takabayashi, J., 'Induced production of extrafloral nectar in intact lima bean plants in response to volatiles from spider mite-infested conspecific plants as a possible indirect defense against spider mites', *Oecologia* 147 (3), (2006): 455–60; https://doi.org/10.1007/s00442-005-0289-8.

40. Harrod Buhner, *ibid.*, pp.126–8

41. Wohlleben, *op. cit.*, p.13

42. Veits, M., *et al.*, 'Flowers respond to pollinator sound within minutes by increasing nectar sugar concentration', *Ecology Letters* 22 (9), (2019): 1,483–92

43. Simard, S., *Finding the Mother Tree: Uncovering the wisdom and intelligence of the forest*, Allen Lane, London, 2021

44. Sheldrake, M., 'Before Roots' in *This Book is a Plant*, Profile Books, London, 2022, pp.9–17

45. Harrod Buhner, *op. cit.*, p.125

46. Simard, *op. cit.*

47. Margulis, L., and Fester, R., *Symbiosis as a Source of Evolutionary Innovation*, MIT Press, Cambridge, MA, 1991

48. Sender, R., Fuchs, S., and Milo, R., 'Revised estimates for the number of human and bacteria cells in the body', PloS Biology, 14 (8), (2016), e1002533; https://doi.org/10.1371/journal.pbio.1002533

49. Sheldrake, M., *Entangled Life: How fungi make our worlds, change our minds, and shape our futures*, Penguin, London, 2021, p.4,

50. Wohlleben, *op. cit.*, p.52

51. Adamatzky, A., 'Language of fungi derived from their electrical spike activity', *Royal Society Open Science* 9, (2022): 4; https://doi.org/10.1098/rsos.211926

52. This also speaks to the metaphysical distinction between 'spirit' and 'soul' in western theology. The 'soul' of a plant, or *philosophical sulphur*, is its more or less unique essential oil, while the 'spirit' is a vitalizing substance common to all plants. *See* Popham, S., *Evolutionary Herbalism: Science, spirituality, and medicine from the heart of nature*, North Atlantic Books, Berkeley, CA, 2019, pp.127–32.

53. Daws, R.E., Timmermann, C., Giribaldi, B., *et al.*, 'Increased global integration in the brain after psilocybin therapy for depression', *Nature Medicine* 28, (2022): 844–51; https://doi.org/10.1038/s41591-022-01744-z

54. Harrod Buhner, S., *Plant Intelligence and the Imaginal Realm: Into the dreaming of Earth*, Bear & Company, Rochester, VT, 2014, pp.200–214

55. *Ibid.*, pp.211–14

56. Ibid.

57. Ibid.

58. Babcock, G.O., 'The Split-Body Problem: Why we need to stop thinking about parents, offspring and sex when we try to understand how life reproduces itself', *Aeon*, April 28, 2022; https://aeon.co/essays/we-need-to-stop-thinking-about-sex-when-it-comes-to-reproduction [accessed February 15, 2023]

59. Crosby, A.W., *The Columbian Exchange: Biological and cultural consequences of 1492*, Greenwood, Westport, CT, 1972

60. Nash, L., 'Beyond Virgin Soils: Disease as environmental history' in *The Oxford Handbook of Environmental History*, ed. A.C. Isenberg, Oxford University Press, Oxford, 2014, pp.76–7

61. Wenner, M., 'Humans carry more bacterial cells than human ones', *Scientific American*, November 30, 2007; https://www.scientificamerican.com/article/strange-but-true-humans-carry-more-bacterial-cells-than-human-ones/ [accessed November 11, 2021]

62. Rybicki, E.P., 'The classification of organisms at the edge of life or problems with virus systematics', *South African Journal of Science* 86, (1990): 182–6

63. NRM, 'Microbiology by numbers', *Nature Reviews Microbiology* 9, (2011): 628; doi: 10.1038/nrmicro2644. PMID: 21961177

64. For more on the origins of viruses, *see* Wessner, D.R., 'The origins of viruses', *Nature Education* 3 (9), (2010): 37.

65. Horie, M., Honda, T., Suzuki, Y., *et al.*, 'Endogenous non-retroviral RNA virus elements in mammalian genomes', *Nature* 463 (7277), (2010): 84–7; doi: 10.1038/nature08695

66. Cited in Zimmer, C., 'Ancient viruses are buried in your DNA', *The New York Times*, Oct. 4, 2017; https://www.nytimes.com/2017/10/04/science/ancient-viruses-dna-genome.html [accessed Sept. 1, 2022]

67. Ibid.

68. Cited in Honigsbaum, M., *The Pandemic Century: A history of global contagion from the Spanish flu to Covid-19*, 2020 edition, Penguin Random House UK, London, 2019, pp.xiv–xv

69. Harrod Buhner, S., *Plant Intelligence and the Imaginal Realm: Into the dreaming of Earth*, Bear & Company, Rochester, VT, 2014, pp.107–109

70. Ibid.

71. Kutter, E., De Vos, D., Gvasalia, G., *et al.*, 'Phage therapy in clinical practice: treatment of human infections', *Current Pharmaceutical Biotechnology* 11 (1), (2010): 69–86. Bacteriophages, viruses that target and 'eat' bacteria, have been used in some parts of the world since the 1920s in the treatment of bacterial infections, and while this practice was eclipsed in the 20th century by the use of antibiotics in western biomedicine, in recent years there has been a resurgence

of interest in the use of 'trained viruses' to treat antibiotic-resistant infections.

72. Honigsbaum, M., *op. cit*, p.xv

Chapter 3: Balance and Imbalance

1. Peoples, H.C., Duda, P., and and Marlowe, F.W., 'Hunter-Gatherers and the Origins of Religion', *Human Nature*, 27, (2016): 261–82; https://doi.org/10.1007/s12110-016-9260-0. *See also* Barnard, A., and Woodburn, J., 'Property, Power and Ideology in Hunter-Gathering Societies: An Introduction,' in *Hunters and Gatherers: Property, power and ideology*, Vol. II, Berg, Oxford, 1988, pp.4–31

2. Tylor, E.B., *Primitive Culture: Researches into the development of mythology, philosophy, religion, art, and custom*, John Murray, London, 1871

3. Latour, B., 'How Not to (De-)animate Nature' in *Facing Gaia: Eight lectures on the new climatic regime*, Polity Press, Cambridge, 2017, p.70

4. Peoples, Duda, and Marlowe, *op. cit.*

5. *Ibid.* Tylor and others postulated that ancestor worship was an early and pervasive element in hunter-gatherer religion, and that this quickly precipitated into a belief in gods. But Peoples *et al.* conclude that both assumptions are ill-founded.

6. Stringer, M.D., 'Rethinking animism: thoughts from the infancy of our discipline', *Journal of the Royal Anthropological Institute* 5 (4), (1999): 541–56

7. Hershkovitz, I., Weber, G.W., Quam, R., *et al.*, 'The earliest modern humans outside Africa', *Science* 359 (6,374), (2018): 456–9. Note that archeological and genetic research forces us to constantly revise the chronology of this model, however, with recent evidence of early Sapiens migrations into Eurasia tracing back as far as 175,000 years.

8. Peoples, Duda, and Marlowe, *op. cit.*, 274–7

9. Nielsen, M., Langley, M.C., Shipton, C., *et al.*, '*Homo neanderthalensis* and the evolutionary origins of ritual in *Homo sapiens*', *Philosophical Transactions of the Royal Society B*, June 29, 2020; http://doi.org/10.1098/rstb.2019.0424

10. Davis, N., '*Homo erectus* may have been a sailor – and able to speak', *Guardian*, 20 Feb. 2018; https://www.theguardian.com/

science/2018/feb/20/homo-erectus-may-have-been-a-sailor-and-able-to-speak [accessed October 5, 2021]

11. Condemi, S., Mazières, S., Faux, P., *et al.*, 'Blood groups of Neandertals and Denisova decrypted', *Plos One* 16 (7), (2021); https://doi.org/10.1371/journal.pone.0254175

12. Jeffrey D., Wall, K.E., and Lohmueller, V.P., 'Detecting ancient admixture and estimating demographic parameters in multiple human populations', *Molecular Biology and Evolution* 26 (8), (2009): 1,823–7; https://doi.org/10.1093/molbev/msp096

13. This is known as 'Dunbar's number', named for British anthropologist Robin Dunbar. For more on this, *see* Hernando, A., Villuendas, D., Vesperinas, C., *et al.*, 'Unravelling the size distribution of social groups with information theory on complex networks', *The European Physical Journal B* 76, (2009): 87–97

14. Harari, Y.N., *Sapiens: A brief history of humankind*, Penguin Random House, London, 2015, pp.22–36

15. *Ibid.*, pp.30–31

16. See Fagan, B., *Cro-Magnon: How the Ice Age gave birth to the first modern humans*, Bloomsbury Press, New York, 2010.

17. Harvey, G., *Animism: Respecting the living world*, Columbia University Press, New York, 2nd ed., 2017, p.xi

18. Gander, K., 'Road project in Iceland delayed to protect "hidden" elves', *Independent*, 23 Dec. 2013; https://www.independent.co.uk/news/world/europe/road-project-in-iceland-delayed-to-protect-hidden-elves-9021768.html [accessed April 2, 2022]

19. Kirk, R., *The Secret Commonwealth of Elves, Fauns and Fairies*, Anodos Books, Dumfries & Galloway, 2018

20. Smith, D.B., 'Mr. Robert Kirk's Note-Book', *The Scottish Historical Review*, 18 (72), (1921): 237–48

21. Fimi, D., *Tolkien, Race and Cultural History: From Fairies to Hobbits*, Palgrave Macmillan, Basingstoke, 2008, pp.29–35

22. Earls, M., 'Happy the Elephant Denied Personhood, Will Stay in Bronx Zoo, Bloomberg Law', *US Law Week*, June 14, 2022; https://news.bloomberglaw.com/us-law-week/happy-the-elephant-denied-personhood-will-remain-in-bronx-zoo [accessed July 17, 2022]

23. *Guardian* staff and agencies, 'Happy the elephant Happy the elephant is not a person, says court in key US animal rights case', *Guardian*, 14 June 2022; https://www.theguardian.com/us-news/2022/jun/14/elephant-person-human-animal-rights-happy [accessed July 17, 2022]

24. Nonhuman Rights Project; https://www.nonhumanrights.org/frequently-asked-questions/ [Accessed July 17, 2022]

25. Ibid.

26. Marder, M., 'Should Animals have Rights?' *Philosopher's Magazine* 62, (2013): 56–7; doi:10.5840/tpm20136293

27. Weeks, L., 'Recognizing the right of plants to evolve', *NPR*, Oct. 26, 2012; https://www.npr.org/2012/10/26/160940869/recognizing-the-right-of-plants-to-evolve?t=1653996829735 [accessed July 25, 2022]

28. CELDF Organizers Markie Miller and Crystal Jankowski have distinguished this approach from extensions of 'personhood' to non-human beings and ecosystems, responding to a prominent critique that such efforts would ultimately pit the personhood of nature against the personhood of humans in developing nations. *See* CELDF, 'Guest Blog: A conversation with the *Guardian*', Dec. 10, 2019; https://celdf.org/2019/12/guest-blog-a-conversation-with-the-guardian/ [accessed March 20, 2022]

29. Articles 71–4, Title II: Rights in the Constitution of the Republic of Ecuador, published in the Official Register on October 20, 2008

30. Ecocide Law, 'Legal Definition and Commentary 2021', *Ecocide Law*, 2021; https://ecocidelaw.com/legal-definition-and-commentary-2021/ [accessed Sept. 4, 2022]

31. Higgins, P., University of Exeter Law School page, 2015; https://law.exeter.ac.uk/cornwall/opportunities/imagine/[accessed Sept. 10, 2022]

32. Lloyd, S.A., Sreedhar, S., 'Hobbes's Moral and Political Philosophy' in *Stanford Encyclopedia of Philosophy*, Stanford University, Stanford, CA, 2018; https://plato.stanford.edu/entries/hobbes-moral/ [accessed May 5, 2022]

33. Hobbes, T., and Curley, E.M., *Leviathan: With selected variants from the Latin edition of 1668*, Hackett Publishing Co., Indianapolis, IN, 1994

34. *Ibid., see* Chapter 13.

35. Rousseau, J-J., *Discourse on the Origin of Inequality*, Hackett Publishing Co., Indianapolis, IN, 1992

36. Graeber, D., and Wengrow, D., *The Dawn of Everything: A new history of humanity*, Farrar, Straus and Giroux, New York, 2021, pp.1–21

37. Burroughs, J., 'The Faith of a Naturalist' in *Accepting the Universe*, Houghton Mifflin, Boston, 1920, pp.88–90

Chapter 4: Natural Hierarchies

1. Hall, M., *Plants as Persons: A philosophical botany*, SUNY Press, Albany, NY, 2011. This section has been greatly influenced by Hall's work.

2. Ruff, C.B., Trinkaus, E., and Holliday, T.W., 'Body Mass and *Encephalization* in Pleistocene *Homo*', *Nature* 387, (1997): 173–6. *See also* Harari, Y.N., *Sapiens*, Penguin Random House, London, 2015, pp.55–6.

3. Mithen, S.J., *After the Ice: A global human history, 20,000–5,000 BC*, Harvard University Press, Cambridge, MA, 2003, pp.29–39

4. Liu, L., Wang, J., Rosenberg, D., *et al.*, 'Fermented beverage and food storage in 13,000 y-old stone mortars at Raqefet Cave, Israel: investigating Natufian ritual feasting', *Journal of Archeological Science: Reports* 21, (2018): 783–93; https://doi.org/10.1016/j.jasrep.2018.08.008

5. Mithen, *op. cit.*, p.50

6. Sweatman, M.B.,'The Younger Dryas impact hypothesis: Review of the impact evidence', *Earth-Science Reviews* 218, (2021):103677; doi:10.1016/J.EARSCIREV.2021.103677

7. Mithen, *op. cit.*, p.48

8. *Ibid.*, pp.48–55

9. Curry, A., 'Last Stand of the Hunter-Gatherers?', *Archeology*, Archeology Institute of America, 2021; https://www.archeology.org/issues/422-2105/features/9591-turkey-gobekli-tepe-hunter-gatherers [accessed July 30, 2022]

10. Cited *ibid.*

11. Harari, Y.N., *Sapiens*, Penguin Random House, London, 2015, pp.101–102

12. Sweatman, M.B., and Tsikritsis, D., 'Decoding Göbekli Tepe with archeoastronomy: what does the fox say?', *Mediterranean Archeology and Archeometry* 17 (1), (2017): 233–50

13. Jones, J., 'Becoming a Centaur', *Aeon*, Jan. 14, 2022; https://aeon.co/essays/horse-human-cooperation-is-a-neurobiological-miracle [accessed August 25, 2022]

14. Abram, D., *The Spell of the Sensuous: Perception and language in a more-than-human world*, Random House, New York, 1997, p.103

15. Ibid.

16. Note that the semi-legendary events of the *Iliad* are traditionally placed in the 12th–13th centuries BCE.

17. Veyne, P., *Did the Greeks Believe in Their Myths? An essay on the constitutive imagination*, trans. P. Wissing, University of Chicago Press, Chicago, IL, 1988

18. Kelley, D.R., *Versions of History: From antiquity to the Enlightenment*, Yale University Press, New Haven, CT, 1991, pp.19–29

19. Abram, *op. cit.*, pp.93–102. The Phoenician script, used by Canaanites in the Levant, ultimately inspired Hebrew and the Brahmi script of ancient India, as well as the Etruscan script found on the Italian peninsula. The Etruscan script went on to form the Roman script (used to write these very words), which now comprises the written standard for over 100 languages to this day. Even the mystical Germanic and Scandinavian runic alphabets, immensely popular in neo-Pagan mysticism, were ultimately derived from Roman and Greek – and thus also Canaanite – scripts.

20. *Ibid.*, pp.108–109

21. *Ibid.*, pp.99–101

22. Ibid.

23. Hall, M., *Plants as Persons*, SUNY Press, Albany, NY, 2011, pp.20–21

24. Plumwood, V., *Feminism and the Mastery of Nature*, Routledge, London, 1993, p.81. *See also* Hall, M., *ibid.*, p.19.

25. Torres, J., 'Plato's Anthropocentrism Reconsidered', *Environmental Ethics* 43 (2), (2021): 119–41

26. In some of Aristotle's writings, he does occasionally acknowledge plants' capacity to experience tactile sensations. In *De Anima*, he describes plants as having a psychic (that is, mental) component

with a sensitivity to touch, as evidenced by their capacity to detect heat and cold. He also argues that the faculty of touch is a direct precursor to the experience of desire, which might lead us to assume that he would view plants as being perceptive entities. However, his formal hierarchy of beings clearly establishes that plants occupy the lowest rung, thereby backgrounding them to a morally inconsiderable position. *See* Hall, M., *op. cit.*, p.26.

27. Lorenz, H., 'Ancient Theories of Soul' in *The Stanford Encyclopedia of Philosophy*, ed. E.N. Zalta, summer 2009; https://plato.stanford.edu/archives/sum2009/entries/ancient-soul/ [accessed Oct. 3, 2021]

28. Cited in Hall, M., *op. cit.*, p.25

29. Ibid.

30. Barnes, J., 'Life and Work' in *The Cambridge Companion to Aristotle*, Cambridge University Press, Cambridge, 1995, p.9

31. Nixey, C., *The Darkening Age: The Christian destruction of the classical world*, Macmillan, London, 2017

32. Cited in Hall, M., *op. cit.*, p.29

33. *Ibid.*, pp.30–35

34. *Ibid.*, p.35

35. *Ibid.*, pp.29–30

36. Pliny the Elder, *Natural History: A selection*, trans. J.F. Healy, Penguin, London, 2004, p.74

37. *Ibid.*, p.286

38. McInerny, R., and O'Callaghan, J., 'Saint Thomas Aquinas' in *The Stanford Encyclopedia of Philosophy*, ed. E.N. Zalta, summer 2018; https://plato.stanford.edu/archives/sum2018/entries/aquinas/ [accessed April 3, 2022]

39. Singer, P., 'Animals' in *The Oxford Companion to Philosophy*, ed. T. Honderich, Oxford University Press, Oxford, 1995, pp.35–6

40. Garber, D., 'Descartes, Mechanics and the Mechanical Philosophy' in *Renaissance and Early Modern Philosophy*, John Wiley and Sons, Oxford, 2002, p.191

41. Pomeroy, R., 'Scientists have learned from cases of animal cruelty', *RealClear Science*, Jan. 23, 2012; https://www.realclearscience.com/blog/2012/01/scientists-can-be-cruel.html [accessed 10 May, 2022].

See also Calvo, Paco, *Planta Sapiens: Unmasking plant intelligence,* The Bridge Street Press, London, 2022, pp.186–7.

42. Descartes, R., *Philosophical Works of Descartes,* trans. E.S. Haldane and G.R.T. Ross, Cambridge University Press, Cambridge, 1967, p.106

43. It should be noted that Descartes was himself deeply influenced by plants, even if he wasn't consciously aware of it. It's said that each day he drank over *five dozen* cups of coffee – a novel commodity in 17th-century Europe. The caffeine in coffee prompts sympathetic nerves near the sinus node of the heart (our natural 'pacemaker') to release norepinephrine, which narrows focus and increases fear and aggression. It also triggers the release of epinephrine in the adrenal glands, which further propels us into a sympathetic 'fight or flight' response. At the same time, excessive use of coffee can cause notable serotonin depletion in the brain, decreasing nonlinear creativity and stifling our sensory experience. This provides some important context for the Cartesian split between *mind* and *body. See* O'Donoghue, S.P., *The Forest Reminds Us Who We Are,* North Atlantic Books, Berkeley, CA, 2021, p.136.

44. Hall, M., *Plants as Persons,* SUNY Press, Albany, NY, 2011, p.49

45. Snobelen, S.D., 'Isaac Newton, heretic: the strategies of a Nicodemite', *British Journal for the History of Science,* 32 (4), (1999): 381

46. Darwin may have managed to nudge science away from strict anthropocentrism in biological sciences, but a zoocentric bias quickly took its place as a result of what James Wandersee and Elisabeth Schussler refer to as *plant blindness* – 'the inability to see or notice plants in one's environment, leading to the inability to recognize the importance of plants in the biosphere and in human affairs.' *See* Wandersee, J.H., and Schussler, E.E., 'Preventing plant blindness', *American Biology Teacher* 61 (2), (1999): 82–6; doi:10.2307/4450624

47. Hayward, J.W., 'Scientific Method and Validation', *Gentle Bridges: Conversations with the Dalai Lama on the sciences of mind,* Shambhala Publications, Boston/London, 2001, p.26

48. *Ibid.,* pp.28–9

49. *Ibid.,* p.29

Chapter 5: Gods and Demons

1. *Thrice-omni* refers to the *omniscience, omnibenevolence,* and *omnipotence* attributed to Yahweh in Abrahamic religious traditions. The incompatibility of these traits with a world that contains 'evil' is the basis of the 'problem of evil' paradox: If a creator god produces evil, then they are not all-loving; if they are incapable of stopping evil, then they are not all-powerful; and if they are unaware of the evil choices that humans might make with their 'free will,' then they cannot be all-knowing. As thrice-omni creators also tend to be perceived as *omnipresent,* this form of theology implies that every time a child is abused, there is an all-knowing, all-powerful, and all-loving god who is fully aware, present, and capable of intervening, but chooses not to.

2. Peoples, H.C., Duda, P., and Marlowe, F.W., 'Hunter-gatherers and the origins of religion', *Human Nature* 27 (2016): 261–82

3. Sample, I., 'Neanderthals built mysterious cave structures 175,000 years ago', *Guardian*, 25 May 2016; https://www.theguardian.com/science/2016/may/25/neanderthals-built-mysterious-cave-structures-175000-years-ago [accessed Oct. 10, 2021]

4. O'Connell, C., *Wild Rituals: 10 lessons animals can teach us about connection, community, and ourselves,* Chronicle PRISM, San Francisco, CA, 2021

5. Harrison, P., *'Religion' and the Religions in the English Enlightenment,* Cambridge University Press, Cambridge, 1990, p.1

6. Ethnologue (n.d.), 'What are the largest language families?'; https://www.ethnologue.com/guides/largest-families. For a full and up-to-date list of Indo-European languages, see https://www.ethnologue.com/subgroups/indo-european [accessed April 23, 2022]

7. For a thorough introduction to Proto-Indo-European, see Anthony, D.W., *The Horse, the Wheel, and Language: How Bronze-Age riders from the Eurasian steppes shaped the modern world,* Princeton University Press, Princeton, NJ/Oxford, 2007. *See also* Mallory, J.P., and Adams, D.Q., *The Oxford Introduction to Proto-Indo-European and the Proto-Indo-European World,* Oxford University Press, Oxford, 2006.

8. There are notably multiple different 'versions' of PIE, based on the research of multiple different philologists.

9. Krzewińska, M., Kılınç, G.M., Juras, *et al.*, 'Ancient genomes suggest the eastern Pontic-Caspian steppe as the source of western Iron Age

nomads', *Science Advances* 4 (10), (2018); doi: https://doi.org/10.1126/sciadv.aat4457

10. Lincoln, B., 'The Indo-European myth of creation', *History of Religions* 15 (2), (1975): 121–45; doi:10.1086/462739

11. Sturluson, S., *The Prose Edda by Snorri Sturluson*, trans. A.G. Brodur, Oxford University Press, Oxford, 1916, pp.20–22

12. Sturluson, S., *The Prose Edda: Also called Snorre's Edda, or The Younger Edda*, trans. and annotated R.B. Anderson, University of Wisconsin, Madison, WI, 1879, pp.64–5

13. Hall, M., *The Imagination of Plants: A book of botanical mythology*, SUNY Press, Albany, NY, 2019, p.30. *See also* White, J., *The Ancient History of the Maori, His Mythology and Traditions*, Vol. 1, George Didsbury, Government Printer, Wellington, NZ, 1887

14. Hall, *ibid.*, pp.3–6

15. The nature of this migration is hotly debated in modern-day India, with many Indian scholars preferring to characterize the Indo-Aryan migration as a relatively peaceful affair. This remains debated in global scholarship, however, with significant evidence pointing to the presence of violence in this period.

16. Anthony, D.W., *The Horse, the Wheel, and Language: How Bronze-Age riders from the Eurasian steppes shaped the modern world*, Oxford University Press, Oxford/Princeton, NJ, 2007, pp.48–50

17. Cited in Hall, *op. cit.*, pp.163–4. This excerpt is taken from Section 184, *Mokshadharma Parva, Santi Parva, The Mahabharata*.

18. Figures like Shiva and Vishnu only rose to prominence as supreme deities in later Indian religious movements that correlate to what are now called 'Hinduism.' Shiva is not present as a distinct god in early Vedic literature, and Vishnu is a minor deity compared to the chief divinities of the Vedic pantheon like Agni and Indra. It should be noted that 'Hinduism' is phenomenally complex. It is not a single religion in the same way that Islam or Christianity are a single religion, even when taking sects into account. It is also not equivalent to Vedism. Hinduism is taxonomically more akin to the notion of an 'Abrahamic' religious worldview underlying Judaism, Christianity, and Islam. There is a degree of shared lineage, but there are ultimately many 'religions' built upon the Hindu worldview, and a wide range of philosophical positions that can be taken. The term 'Hinduism,' like the name 'India,' was notably introduced by the British. Modern Hindus usually identify themselves as practitioners

of *Sanātana Dharma*, but may otherwise identify more specifically as *Shaivites*, *Vaishnavites*, or adherents of another specific sect.

19. Kuiper, F.B.J., 'The Basic Concept of Vedic Religion', *History of Religion* 15, (1975): 108–12. *See also* Parpola, A., *The Roots of Hinduism: The early Aryans and the Indus civilization*, Oxford University Press, Oxford, 2015, pp.66–7, 82–109.

20. Gignoux, P., 'Hell in Zoroastrianism', *Encyclopedia Iranica* XII (2), (2003): 154–6; https://www.iranicaonline.org/articles/hell-i [accessed March 17, 2022]

21. Wade, N., *The Faith Instinct: How religion evolved and why it endures*, Penguin, New York, 2009, p.150

22. Ibid.

23. *Ibid.*, pp.150–54. Note that the Deuteronomic Histories include the Books of Joshua, Judges, 1 and 2 Samuel, and 1 and 2 Kings. The 18th–19th-century German theologian Wilhelm de Wette concluded that these texts were all discovered or composed during the renovation of 622BCE.

24. *Ibid.*, p.155

25. Ibid.

26. Ibid.

27. *Ibid.*, p.156

28. Ibid.

29. It should further be noted that of the 13 Pauline epistles, only seven have been confirmed to be authentically authored by Paul himself.

30. We generally think of 'world religion' as being synonymous with 'the religions of the world,' but in fact there are only a handful of religious traditions that have gained a global following. These are, in more or less chronological order, Zoroastrianism, Buddhism, Christianity, Manichaeism, and Islam. These are 'world religions' in the truest sense.

31. Nixey, C., *The Darkening Age: The Christian destruction of the classical world*, Macmillan, London, 2017, pp.8–9

32. *Ibid.*, pp.xxvii–xxxix

33. *See* Wade, N., *The Faith Instinct: How religion evolved and why it endures*, Penguin, New York, 2009, pp.178–84. Revisionist scholars have come to question whether the name Muhammad, meaning 'the chosen one,' is a personal name at all.

34. Genesis 1:26–31, ESV

35. White, L., 'The Historical roots of our ecological crisis', *Science* 155, (1967): 1,203–207

36. Denova, R., 'The Origin of Satan', *World History Encyclopedia*, 2021; https://www.worldhistory.org/article/1685/the-origin-of-satan/ [accessed Oct. 20, 2021]

37. Kelly, H.A., *Satan: A biography*, Cambridge University Press, Cambridge, 2006, pp.1–13, 28–9

38. Saint Augustine, *'Psalms 73–98', Exposition on the Psalms, Vol. 4*, trans. M. Boulding, Augustinian Heritage Institute, New York, 2002, p.94. See also Nixey, C., *The Darkening Age: The Christian destruction of the classical world*, Macmillan, London, 2017, p.18

39. Broedel, H.P., *The* Malleus Maleficarum *and the Construction of Witchcraft: Theology and popular belief*, Manchester University Press, Manchester, 2003, p.1

40. *Ibid.*, p.17

41. *Ibid.*, pp.17–18

42. *Ibid.*, pp.20–22

43. *Ibid.*, pp.29–31

44. Mackay, C.S., *The Hammer of Witches: A complete translation of the* Malleus Maleficarum, Cambridge University Press, Cambridge, 2009, p.19

45. Owens, Y., 'The saturnine history of Jews and witches', *Preternature: Critical and historical studies on the preternatural* 3 (1), (2014): 56–84; JSTOR, https://doi.org/10.5325/preternature.3.1.0056

46. Broedel, *op. cit.*, p.2

47. Levack, B.P., *The Oxford Handbook of Witchcraft in Early Modern Europe and Colonial America*, Oxford University Press, Oxford, 2013, pp.75–6

Chapter 6: Provocation

1. Adam, D., '15 million people have died in the pandemic, WHO says', *Nature* 605, (2022): 206; doi: https://doi.org/10.1038/d41586-022-01245-6

2. Watson, O.J., Barnsley, G., Toor, J., *et al.*, 'Global impact of the first year of COVID-19 vaccination: a mathematical modelling study,' *The Lancet Infectious Diseases* 22 (9), (2022); doi https://doi.org/10.1016/S1473-3099(22)00320-6

3. Sholtis, B., 'When COVID deaths are dismissed or stigmatized, grief is mixed with shame and anger', *NPR*, Aug. 30, 2021; https://www.npr.org/sections/health-shots/2021/08/30/1011785899/when-covid-deaths-are-dismissed-or-stigmatized-grief-is-mixed-with-shame-and-ang?t=1661861693246 [accessed Aug. 29, 2022]

4. Dobson, A.P., and Carper, E.R., 'Infectious diseases and human population history', *Bioscience* 46, (1996): 115–26

5. 'Sharing notes' of course requires things to be written down, even though a great deal of information can also be transmitted orally or through demonstration. Writing was a profoundly useful innovation in the history of science, in no small part because it allowed for information to be transmitted through space and time without a highly trained intermediary. Knowledge became arguably more cumulative, but also more suited for critical reflection and engagement.

6. Gyu thog yon tan mgon po, *gDon nad gso ba bcos sgrig ma*, ed. and annotated by Bya mdo klu byams rgyal, Mi rigs dpe skrun khang, Beijing, 2019 [BDRC bdr:MW8LS68144]

7. Krug, A., 'Buddhist medical demonology in the sutra of the seven buddhas', *Religions* 10 (4), (2019); https://doi.org/10.3390/rel10040255

8. Saxer, M., 'The Journeys of Tibetan Medicine' in *Bodies in Balance: The art of Tibetan Medicine*, Rubin Museum of Art and University of Washington Press, New York/Seattle, 2014, pp.246–56

9. *Ibid. See also* Martin Saxer's 2005 film, *Journeys with Tibetan Medicine*, available online at https://vimeo.com/122821844.

10. Honigsbaum, M., *The Pandemic Century*, 2020 edition, p.xv

11. Ibid.

12. For an overview of the history of vaccination, *see* Riedel, S., 'Edward Jenner and the history of smallpox and vaccination', *Baylor University Medical Center (Proceedings)* 18 (1), (2005): 21–5.

13. *Ibid.* Rev. Mather (1663–1728) and Dr Zabdiel Boylston (1679–1766) popularized the practice of variolation in the colonies, albeit with significant resistance. Dr Mather's home was bombed during the height of the smallpox epidemic of 1721. But ultimately, their success with variolation through this epidemic – demonstrating a reduction in mortality from 14 per cent to 2 per cent – helped to bolster support for the practice in Europe.

14. 'Sa gnyan rlog nas spang tha zhing du 'dru | chu gnyan dkrugs nas ne'u gsing rdzing du bskyil | shing gnyan gcod cing rdo gnyan rtsa ba 'degs | mi gtsang thab gzhob shan dmar 'jol nyog spyod | ban bon nus pa sgrub pa'i long med nas | gnyan sa dkrugs pas dgra rnams thul la re.' See Mgon po, gDon nad gso ba bcos sgrig ma, ed. Bya mdo klu byams rgyal, Mi rigs dpe skrun khang, Beijing, 2019, p.144.

15. Ibid.

16. Kocurek, J., 'Tree beings in Tibet: contemporary popular concepts of *kLu* and *gNyan* as a result of ecological change', *Journal of Ethnology and Folkloristics* 7 (1), (2013): 26

17. Ibid.

18. *Ibid.*, pp.26–7

19. *Ibid.*, p.28

20. *Ibid.*, pp.52–69. Note that this conceptualization of spirit illness is not Yuthok's invention, but was adapted from the Ayurvedic tradition. While his treatment of 'possession' is closely derived from Vagbhata's work, his approach to the ecological dynamics of provocation are quite original in many respects. *See* Yang Ga, 'Sources for the writing of the *rgyud bzhi*, Tibetan medical classic', doctoral dissertation, Harvard University, Cambridge, MA, 2010, pp.229–32.

21. Coope, J., 'On the need for an ecologically dimensioned medical humanities', *Medical Humanities* 47 (1), (2021): Introduction

Chapter 7: Liberating the Unseen

1. Levman, B.G., 'Cultural remnants of the indigenous peoples in the Buddhist scriptures', *Buddhist Studies Review* 30 (2), (2014): 145–80; doi:10.1558/bsrv.v30i2.145

2. Ibid.

3. *Gaina Sutras, Part 1: The Âkârânga Sùtra, the Kalpa Sûtra*, trans. H. Jacobi, Clarendon Press, Oxford, 1884, pp.80–81. *See also* Hall, M., *Imagination of Plants: A book of botanical mythology*, SUNY Press, Albany, NY, 2019, p.200.

4. This is not to say that *all* Buddhists have *always* been exclusively 'nonviolent.' Even today, we need look no further than Myanmar to see that this is not the case. There are also many examples of ritual 'violence' (against unseen beings) in the Tibetan Vajrayana tradition. But as far as the sphere of moral consideration is concerned, the Buddha's view was that the ethical treatment of non-human beings *should* be a foundational feature of a noble philosophical paradigm.

5. Ham, H.S., 'Manipulating the memory of meat-eating: reading the Lankāvatāra's strategy of introducing vegetarianism to Buddhism', *Journal of Indian Philosophy* 47, (2019): 133–53; https://doi.org/10.1007/s10781-019-09382-5

6. Hall, M., *Plants as Persons: A philosophical botany*, SUNY Press, Albany, NY, 2011, pp.87–8

7. Cited in Hall, M., *The Imagination of Plants: A book of botanical mythology*, SUNY Press, Albany, NY, 2019, p.140

8. Cited in Hall, M., *Plants as Persons, op. cit.*, p.97

9. Herold, A.F., *The Life of the Buddha*, trans. P.C. Blum, A. and C. Boni, New York, 1927, pp.87–8. *See also* Hall, M., *The Imagination of Plants, op. cit.*

10. Herold, *ibid.*

11. Haberman, D.L., *People Trees: Worship of trees in northern India*, Oxford University Press, Oxford, 2013, pp.49–59

12. Cunningham, A., *Mahâbodhi, or the great Buddhist temple under the bodhi tree at Buddha-Gaya*, W.H. Allen & Co., London, 1892, pp.30–31

13. Forbes, S., 'The oldest historical tree in the world', *Medium*, Dec. 26, 2016; https://medium.com/@StephenJForbes/the-oldest-historical-tree-in-the-world-566fcee32605 [accessed May 11, 2022]

14. Cunningham, A., *op. cit.*, pp.30–31

15. Cited *ibid.*, title page

16. See Garling, W., *Stars at Dawn: Forgotten stories of the women in the Buddha's life*, Shambhala Publications, Boulder, CO, 2016.

17. Rockhill, W.W., *Life of the Buddha and the Early History of His Order: Derived from Tibetan works in the Bkah-hgyur [sic] and Bstan-hgyur [sic]*, Trübner & Co., London, 1884, pp.80–81

18. Levman, B.G., *Pāli and Buddhism: Language and lineage*, Cambridge Scholars Publishing, Newcastle upon Tyne, 2021, p.99, footnote 155. Note that the species of the trees at the Buddha's birth is contested in later retellings of the story, but the earliest layers clearly refer to them as *Shorea robusta*.

19. Levman, B.G., 'Cultural Remnants of the indigenous peoples in the Buddhist scriptures', *Buddhist Studies Review* 30 (2), (2014): 145–80

20. Lingpa, J., and Kangyur Rinpoche, *Treasury of Precious Qualities: Book One*, Shambhala Publications, Boulder, CO, 2010, pp.76–7

21. *Ibid.*, pp.77–8

22. *Ibid.*, p.28

23. Hall, M., *The Imagination of Plants: A book of botanical mythology*, SUNY Press, Albany, NY, 2019, p.151. Many similarities can be found between the *nāgas* of India and the *nymphs* of ancient Greece. Both are associated with topographic features and, most particularly, bodies of water. By extension, both are also closely associated with vegetation, and most of all trees. In Greece, *dryads* (tree nymphs), *oreads* (mountain nymphs), and *naiads* (water spring nymphs) are all seen to be 'kinds' of nymphs. In India, *nāgas* are often considered to live in caste-based societies, with *brahmin nāgas* viewed as wizened holders of knowledge, and lower-caste *nāgas* perceived as dangerous, bloodthirsty spirits that should be avoided at all costs. In Tibet, the *klu* (a traditional spirit class assimilated to *nāga* in the Buddhist era) can further be classified through hybrid categories formed with other beings – i.e. the *klu gnyan, klu srin, klu btsan*, etc.

24. In the *Oral Instruction Tantra* (*Man ngag rgyud*), Yuthok dedicates an entire chapter of his treatment on *provocation disorders* to diseases caused by *nāgas* (Tib. *klu*). These disorders tend to affect the skin and lymphatic system, with leprosy being their most serious manifestation. Yuthok's overview of this topic is notably not derived from any known Ayurvedic source materials (which do not generally

associate *nāgas* with such diseases), nor from earlier Tibetan syntheses like the *Lunar King*. It appears that his understanding of the particular dynamics of *nāga*-related illness may have come from indigenous Tibetan sources, potentially through family lineages outside the mainstream medical milieux. *See* Yang Ga, 'Sources for the writing of the *rgyud bzhi*, Tibetan medical classic', doctoral dissertation, Harvard University, Cambridge, MA, 2010, pp.231–2.

25. In the Epic of King Gesar, the legendary king is believed to have been born from a human father and *nāga* mother. *See* Kornman, R. (trans.), Chonam, L. (trans.), Khandro, S. (trans.), *et al.*, *The Epic of Gesar of Ling: Gesar's magical birth, early years, and coronation as king*, Shambhala Publications, Boulder, CO, 2015.

26. Larson, J., *Greek Nymphs: Myth, cult, lore*, Oxford University Press, Oxford, 2001, p.10. *See also* Hall, M., *The Imagination of Plants, op. cit.*, p.151.

27. The rain-summoning ritual from the *Precious Garland* (*Rin chen 'phreng ba*) *Chöd* collection involves a detailed process of subjugating rain-blocking *nāgas* in order to control the weather. *See* gNas mdo karma chags med, *gCod lugs char 'bod snyan rgyud yi ger bkod par 'don bsgom phyag len dang bcas pa mthong bas don gsal bzhugs so*; BDRC bdr:MW0LULDC313476

28. This is a common understanding even in Tibetan Buddhism, though the indigenous *klu* of Tibet likely didn't follow such paradigms. It's interesting to note similarities with the *húldufolk* of Iceland, who are often characterized as being 'Christian,' even constructing cathedrals within the natural landscape.

29. Levman, B.G., 'Cultural remnants of the indigenous peoples in the Buddhist scriptures', *Buddhist Studies Review* 30 (2), (2014): 145–80; doi:10.1558/bsrv.v30i2.145

30. In the Pali *Sānu Sutta*, a *yakṣini* possesses a monk in order to prevent him from abandoning his spiritual path and proceeds to bestow teachings upon his mother.

31. This concept of tantra as a process of 'undermining' our afflictive tendencies is derived from a lecture delivered by Lama Sarah Harding at Naropa University in 2009.

32. Many Tibetan words are rendered as 'demons,' including *'dre* and *gdon*. *'Dre* is used more generally to refer to malefic entities, the latter, which I translate as 'provocation' or 'provocateur,' is used to refer to a broader range of spirits as well as the human afflictions

that they cause. *Bdud* is used to render Sanskrit *māra* in Tibetan, which is also occasionally used to refer to malevolent supernatural forces (sometimes translated into English as 'devils'), but is more frequently applied in a philosophical context to refer to any phenomenon or experience that hinders our liberation.

33. While some will shudder at my use of this term, 'wizard' is closely cognate with Sanskrit *vidhyadhara*, or 'wisdom-holder,' and carries much of the same semantic *value* in practice.

34. The placation of local land spirits preceding construction was once a common practice across the ancient world, including in Europe, where *augurs* responsible for interpreting geomantic omens would be regularly consulted before doing anything that could disrupt the local spirit ecosystem. Our verb *to in*-augur-*ate* (Latin *inaugurare*) is derived from just such a process, known as *augury*. These figures weren't seen as *fortune-tellers*, but as interpreters of the wills and wishes of unseen beings. Rather than reading cards or lines on a person's hand, augurs relied on the signs of nature itself, namely the behaviors and flight patterns of birds. There's evidence that forms of bird augury were also practiced in the Tibetan world, and many patterns used in Sino-Tibetan elemental astrology are called *cha-lam*, literally 'the bird's [flight] path.' This is also used in the medical tradition, for instance in the *Lunar King*. Chu'i bya lam dus kyi 'khor lo (le'u 12, *sMan dpyad zla ba'i rgyal po*, Mi rigs dpe skrun khang, Beijing, 2006, pp.53–5

35. ma gcig lab sgron, *Machik's Complete Explanation: Clarifying the meaning of Chöd*, trans. S. Harding, Snow Lion Publications, Ithaca, NY, 2013, pp.93–7. It's notable that most of Machig's male contemporaries took a very different approach, and asserted their often original works as the 'revealed' teachings of Indian masters from the past. Machig's does claim to have been an Indian yogi in her past life, but she otherwise doesn't seek to establish an Indian authority for her system.

36. Allione, T., *Women of Wisdom*, revised edition, Snow Lion Publications, Ithaca, NY, 2000, pp.183–4

37. ma gcig lab sgron, *op. cit.*, pp.205–207

38. *Ibid.*, p.117. Harding uses 'devil' to translate *bdud*, which is the Tibetan term for *māra*. This is a standard definition for *māra* in Buddhist philosophy. There is no sense that this *māra* is an agentive personification of evil, but rather a kind of representation of all the forces that hinder our experience of liberation.

39. Allione, T., *Feeding Your Demons: Ancient wisdom for resolving inner conflict*, Little, Brown, New York, 2008. This is a phenomenally useful technique which helped me significantly through my formative teen years, and there is even some notable scientific research to support its efficacy as a therapeutic process.

40. *Ibid.*, ebook version, pp.53–5

Chapter 8: Entering the Perilous Realm

1. Segal, R.A., *Myth: A very short introduction*, Oxford University Press, Oxford, 2004, p.5. *See also* Hall, M., *Imagination of Plants: A book of botanical mythology*, SUNY Press, Albany, NY, 2019, pp.xxi–xxiii

2. Campbell, J., *The Hero with a Thousand Faces*, Fontana Press, London, 1993, p.5. *See also* Hall, *ibid.*, p. xxiii.

3. Mark, J.J., 'Mythology' in *World History Encyclopedia*, 2018; https://www.worldhistory.org/mythology/ [accessed October 12, 2021]

4. Harari, Y.N., *21 Lessons for the 21st Century*, Penguin Random House, London, 2019, p.285

5. *Ibid.*, pp.285–6

6. *Ibid.*, p.289

7. Le Guin, U.K., 'The critics, the monsters, and the fantasists', *The Wordsworth Circle* 38 (1/2), (2007): 87

8. *See* Ulstein, G., 'Hobbits, Ents, and dæmons: ecocritical thought embodied in the fantastic', *Fafnir – Nordic Journal of Science Fiction and Fantasy Research* 2 (4), (2015): 7–17

9. Tolkien, J.R.R., *On Fairy-Stories*, expanded edition, with commentary and notes, HarperCollins, London, 2014, pp.31–2

10. *Ibid.*, pp.68–9

11. *See,* for instance, Letter 203 (written to Herbert Schiro) in Tolkien, J.R.R., *The Letters of J.R.R. Tolkien*, HarperCollins, London, 2006, p.262.

12. Tolkien, J.R.R., *On Fairy-Stories, op. cit.*, p.36

13. *Ibid.*, p.28

14. Chodar, T., Schneider, S. (trans.), and Allione, L.T., *Luminous Moonlight: The biography of Do Dasal Wangmo*, Tara Mandala, Pagosa Springs, 2013

15. The famed *Tibetan Book of the Dead* (AKA the *Bardo Thödrol*) is a *terma* text, as are most works pertaining to the Dzogchen tradition (the *Great Perfection*). Of the latter, the Seventeen Dzogchen Tantras provide some of most lucid and revered explanations on the nature of reality found in the Buddhist corpus.

16. Hutton, C.M., *Linguistics and the Third Reich: Mother-tongue fascism, race, and the science of language*, Routledge, London/New York, 1999, pp.3–13

17. Cited in Scull, C., and Hammond, W.G., 'Northernness' in *The J.R.R. Tolkien Companion and Guide: Reader's Guide*, Part II, HarperCollins, London, 2017, p.862

18. Fimi, D., *Tolkien, Race and Cultural History: From Fairies to Hobbits*, Palgrave Macmillan, Basingstoke, 2008, pp.51–65

19. Cited in Tolkien, J.R.R., and Tolkien, C., 'The Notion Club Papers' in *Sauron Defeated*, HarperCollins, London, 2017, p.236

20. *Ibid.* This is extracted from 'The Notion Club Papers', where Tolkien's autobiographical experience is given to the character Arundel Lowdham, a fictional 20th-century member of the fictional Oxford literary group the Notion Club.

21. Carpenter, H., *J.R.R. Tolkien: A biography*, HarperCollins, London, 2016, p.102

22. Lewis, C.S., *Lewis: Collected Letters, Volume 1: Family Letters. 1905–1931*, ed. W. Hooper, HarperCollins, London, 2000, p.909

23. *Ibid.*, p.129

24. Tolkien, J.R.R., 'Letter 257: to Christopher Bretherton', *The Letters of J.R.R. Tolkien*, HarperCollins, London, 2006, p.347

25. *See* Scull, C., and Hammond, W.G., 'Atlantis' in *The J.R.R. Tolkien Companion and Guide*, revised and expanded edition, HarperCollins, London, 2017, pp.81–4; *also* Tolkien, J.R.R., and Sibley, B. (ed.), *The Fall of Númenor*, HarperCollins, London, 2022, pp.xvi–xxiv.

26. Tolkien, J.R.R., 'Letter 257: to Christopher Bretherton', *The Letters of J.R.R. Tolkien., op. cit.*, p.347

27. Tolkien, J.R.R., *In Their Own Words: British Authors*, BBC interview, 1968

28. I've heard this anecdote from multiple sources, but have yet to find any recorded or written evidence to substantiate its validity or provide context for the statement. One account came from a (now-deceased) western practitioner who had spent a long time in Trungpa's community.

29. Tolkien, J.R.R., *On Fairy-Stories, op. cit.,* pp.65–6

30. Ibid.

31. Tolkien, J.R.R., 'Letter 165 to Houghton Mifflin Co.' in *The Letters of J.R.R. Tolkien*, HarperCollins, London, 2006, p.220

32. *See* Judd, W., and Judd, G.A., *Flora of Middle-Earth*, Oxford University Press, Oxford, 2017.

33. Tolkien, J.R.R., 'Letter 339 to the Editor of *The Daily Telegraph*' in *The Letters of J.R.R. Tolkien*, HarperCollins, London, 2006, pp.419–20

34. Tolkien, J.R.R., and Tolkien, C. (ed.), 'Myths Transformed' in *Morgoth's Ring*, HarperCollins, London, 1993; 2017 edition, pp.369–436

35. Tolkien, J.R.R., *On Fairy-Stories*, expanded edition, with commentary and notes, HarperCollins, London, 2014, p.65

36. *Ibid.,* p.66

37. *Ibid.,* p.67

38. *Ibid.,* p.68

39. *Ibid.,* p.69

40. *Ibid.,* p.153

41. Plumwood, V., 'Nature in the active voice', *Australian Humanities Review* 46, (2009): 127–8

Chapter 9: Methods of Treatment

1. Plumwood, V., *Feminism and the Mastery of Nature*, Routledge, London, 1993, pp.166–71

2. Tolkien, J.R.R., *The Lord of the Rings*, HarperCollins, London, 1994 edition, p.861

3. It's worth noting that, in the Tibetan tradition, there is a practice known as *bcud len* (*chülen*), in which an accomplished practitioner is believed to be able to sustain themselves through the transmutation of the five elements alone. Some great adepts claim to be able to fast indefinitely with these techniques, which in some forms resemble ideas like 'breatharianism.' But conceptions of light are often integral to this process, as well as in Taoist traditions.

4. Hall, M., *Imagination of Plants: A book of botanical mythology*, SUNY Press, Albany, NY, 2019, p.199

5. *Ibid.*, p.198

6. Descola, P., *Beyond Nature and Culture*, University of Chicago Press, Chicago, 2013, pp.15–16

7. Holmes, J., 'Losing 25,000 to hunger every day', *UN Chronicle* (n.d.); https://www.un.org/en/chronicle/article/losing-25000-hunger-every-day

8. Ritchie, H., 'Half of the world's habitable land is used for agriculture', *World in Data*, Nov. 11, 2019; https://ourworldindata.org/global-land-for-agriculture [accessed July 20, 2022]

9. Ritchie, H., 'How much of the world's land would we need in order to feed the global population with the average diet of a given country?'; *World in Data*, Oct. 3, 2017; https://ourworldindata.org/agricultural-land-by-gloabla-diets [accessed July 20, 2022]

10. Woodhall, A., 'Addressing Anthropocentrism in Nonhuman Ethics: Evolution, morality, and nonhuman beings', doctoral dissertation, University of Birmingham Dept. of Philosophy, 2016

11. This is paraphrased from teachings given by Drubpön Lama Karma Jñana during a Rinchen Trengwa retreat at Tara Mandala retreat center in the winter of 2017.

12. Nixon, D., 'The body as mediator: the phenomenology of Maurice Merleau-Ponty entwines us, via our own beating, pulsing, living bodies, in the lives of others', *Aeon*, Dec. 7, 2020; https://aeon.co/essays/the-phenomenology-of-merleau-ponty-and-embodiment-in-the-world [accessed Aug. 12, 2022]

13. Campbell, J., 'The Wisdom of Joseph Campbell', interview with Michael Toms, New Dimensions Radio, 1991; https://www.jcf.org/works/quote/ritual-is-the-enactment/

14. Hayden, L., 'I'm beginning a journey': the inside story of Lorde's surprise mini-album in *te reo Māori'*, *The Spinoff*, Sept. 9, 2021; https://thespinoff.co.nz/atea/09-09-2021/lorde-interview-maori-lyrics-solar-power [Accessed June 5, 2022]

15. Lavin, W., 'Lorde tells fans her album is "going to take some time" after death of her dog Pearl', *NME*, 2 Nov., 2019; https://www.nme.com/news/music/lorde-tells-fans-album-going-take-time-death-dog-2563977 [accessed June 5, 2022]

16. Group Think, 'Lorde's *Te Ao Mārama*: Behind the songs', *The Spinoff*, Sept. 9, 2021; https://thespinoff.co.nz/atea/09-09-2021/lordes-te-ao-marama-behind-the-songs [accessed June 5, 2022]

17. Brown, H., 'Lorde review, *Solar Power*: Disappointing, detached and sun-bleached of melody', *Independent*, 19 Aug., 2021; https://www.independent.co.uk/arts-entertainment/music/reviews/lorde-review-solar-power-b1904612.html [accessed June 5, 2022]

18. Kornhaber, S., 'The pop star who's redefining the word "basic"', *Atlantic*, Aug. 19, 2021; https://www.theatlantic.com/culture/archive/2021/08/lorde-solar-power-review/619808/?utm_source=feed [accessed June 5, 2022]

19. Daly, R., 'Lorde – *Solar Power* review: a dazzling hat-trick from a master of her craft', *NME*, 20 Aug., 2021; https://www.nme.com/reviews/album/lorde-solar-power-review3023597?utm_source=rss&utm_medium=rss&utm_campaign=lorde-solar-power-review [accessed June 5, 2022]

Conclusion: Learning to See

1. Coope, J., 'On the need for an ecologically dimensioned medical humanities', *Medical Humanities* 47 (1), (2021): 123–7

2. Graeber, D., and Wengrow, D., *The Dawn of Everything: A new history of humanity*, Farrar, Straus and Giroux, New York, 2021, p.9

BIBLIOGRAPHY

David Abram, *Becoming Animal*, Vintage Books, New York, 2011

——, *The Spell of the Sensuous: Perception and language in a more-than-human world*, Random House, New York, 1997

David Adam, '15 million people have died in the pandemic, WHO says', *Nature* 605 (206), (2022)

Andrew Adamatzky, 'Language of fungi derived from their electrical spiking activity', *Royal Society Open Science* 9 (4), (2022)

Tsultrim Allione, *Women of Wisdom*, Snow Lion Publications, Ithaca, NY, 2000

——, *Feeding Your Demons: Ancient wisdom for resolving inner conflict*, Little, Brown, New York, 2008

Rosamunde Almond, Monique Grooten, and Tanya Petersen, 'Living Planet Report 2020: Bending the curve of biodiversity loss', World Wildlife Fund (WWF), Gland, Switzerland, 2020

David W. Anthony, *The Horse, the Wheel, and Language: How Bronze-Age riders from the Eurasian steppes shaped the modern world*, Princeton University Press, Princeton, NJ, and Oxford, 2007

Gunnar O. Babcock, 'The Split-Body Problem: Why we need to stop thinking about parents, offspring and sex when we try to understand how life reproduces itself', *Aeon*, April 28, 2022; https://aeon.co/

essays/we-need-to-stop-thinking-about-sex-when-it-comes-to-reproduction [accessed May 20, 2022]

František Baluška, Dieter Volkmann, Andrej Hlavacka, Stefano Mancuso, *et al.*, *Communication in Plants: Neuronal aspects of plant life*, eds František Baluška, Stefano Mancuso and Dieter Volkmann, Springer-Verlag, Berlin, 2006

Alan Barnard and James Woodburn, 'Property, Power and Ideology in Hunter-Gathering Societies: An introduction' in *Hunters and Gatherers: Property, power and ideology*, Vol. 2, eds Tim Ingold, David Riches and James Woodburn, Berg, Oxford, 1988, pp.4–31

Jonathan Barnes, 'Life and Work' in *The Cambridge Companion to Aristotle*, Cambridge University Press, Cambridge, 1995

Stephen Batchelor, *After Buddhism: Rethinking the dharma for a secular age*, Yale University Press, New Haven, CT, 2015

Wolfgang Behringer, *Witches and Witch-hunts: A global history*, Polity Press, Cambridge, 2004

Christopher Bell, *Tibetan Demonology*, Cambridge University Press, Cambridge, 2020

Diana Beresford-Kroeger, *To Speak for the Trees: My life's journey from ancient Celtic wisdom to a healing vision of the forest*, Random House Canada, Toronto, 2019

Sharon Blackie, *The Enchanted Life*, September Publishing, Tewkesbury, 2018

Laura R. Botigué, Shiya Song, Amelie Scheu, *et al.*, 'Ancient European dog genomes reveal continuity since the Early Neolithic', *Nature Communications* 8, (2017) 16082

Eric D. Brenner, Rainer Stahlberg, Stefano Mancuso, *et al.*, 'Plant neurobiology: an integrated view of plant signalling', *TRENDS in Plant Science* 11 (8), (2006): 413–19

Hans Peter Broedel, *The* Malleus Maleficarum *and the Construction of Witchcraft: Theology and popular belief*, Manchester University Press, Manchester, 2003

Stephen Harrod Buhner, *Plant Intelligence and the Imaginal Realm: Into the dreaming of Earth*, Bear & Company, Rochester, VT, 2014

John Burroughs, 'The Faith of a Naturalist' in *Accepting the Universe*, Houghton Mifflin, Boston, 1920

Paco Calvo, *Planta Sapiens: Unmasking plant intelligence*, The Bridge Street Press, London, 2022

Joseph Campbell, *The Hero with a Thousand Faces*, Fontana Press, London, 1993

——, *The Wisdom of Joseph Campbell*, interview with Michael Toms, New Dimensions Radio, 1991; https://www.jcf.org/works/quote/ritual-is-the-enactment/

Humphrey Carpenter, *J.R.R. Tolkien: A biography*, HarperCollins, London, 2016

CELDF (Community Environmental Legal Defense Fund), 'Guest Blog: A conversation twith the *Guardian*', CELDF, Dec. 10, 2019; https://celdf.org/2019/12/guest-blog-a-conversation-with-the-guardian/[accessed March 20, 2022]

Dipesh Chakrabarty, 'The climate of history: four theses', *Critical Inquiry* 35, (2009): 197–222

Thubten Chodar and Lama Tsultrim Allione, *Luminous Moonlight: The biography of Do Dasal Wangmo*, trans. Sarah Schneider, Tara Mandala, Pagosa Springs, 2013

Yasuyuki Choh and Soichi Kugimiya, 'Induced production of extrafloral nectar in intact lima plants in response to volatiles from spider mite-infested conspecific plants as a possible indirect defense against spider mites', *Oecologia* 147 (3), (2006): 455–60

Terry Clifford, *Tibetan Buddhist Psychiatry: The diamond healing*, Samuel Weiser, York Beach, ME, 1984

Silvana Condemi, Stépane Mazières, Pierre Faux, *et al.*, 'Blood groups of Neandertals and Denisova decrypted', *Plos One* 16 (7), (2021)

Jonathan Coope, 'On the need for an ecologically dimensioned medical humanities', *Medical Humanities* 47 (1), (2021): 123–7

Alfred W. Crosby, *The Columbian Exchange*, Greenwood, Westport, CT, 1972

Alexander Cunningham, *Mahâbodhi, or the Great Buddhist Temple under the Bodhi Tree at Buddha-Gaya*, W.H. Allen & Co., London, 1892

Andrew Curry, 'Last stand of the hunter-gatherers?', *Archeology*, Archeological Institute of America, June 2021; https://www.archeology.org/issues/422-2105/features/9591-turkey-gobekli-tepe-hunter-gatherers [accessed July 30, 2021]

Patrick Curry, *Defending Middle-Earth: Tolkien: Myth and Modernity*, St. Martin's Press, New York, 1997

Rhian Daly, 'Lorde – *Solar Power* review: a dazzling hat-trick from a master of her craft', *NME*, 20 Aug., 2021; https://www.nme.com/reviews/album/lorde-solar-power-review-3023597?utm_source=rss&utm_medium=rss&utm_campaign=lorde-solar-power-review [accessed June 5, 2022]

Nicola Davis, '*Homo erectus* may have been a sailor – and able to speak', *Guardian*, 20 Feb., 2018; https://www.theguardian.com/science/2018/feb/20/homo-erectus-may-have-been-a-sailor-and-able-to-speak [accessed Oct. 5, 2021]

Richard E. Daws, Christopher Timmermann, Bruna Giribaldi, *et al.*, 'Increased global integration in the brain after psilocybin therapy for depression', *Nature Medicine* 28, (2022): 844–51

Rebecca Denova, 'The Origin of Satan', *World History Encyclopedia*, Feb. 18, 2021; https://www.worldhistory.org/article/1685/the-origin-of-satan/ [accessed Oct. 20, 2021]

René Descartes, *Philosophical Works of Descartes*, trans. Elizabeth S. Haldane and G.R.T. Ross, Cambridge University Press, Cambridge, 1967

Philippe Descola, *Beyond Nature and Culture*, University of Chicago Press, Chicago, 2013

Andrew P. Dobson and E. Robin Carper, 'Infectious diseases and human population history', *Bioscience* 46, (1996): 115–26

Maya Earls, 'Happy the elephant denied personhood, will stay in Bronx Zoo', *Bloomberg Law*, June 14, 2022; https://news.bloomberglaw.com/us-law-week/happy-the-elephant-denied-personanhood-will-remain-in-bronz-zoo [accessed July 17, 2022]

Ethnologue, 'What are the largest language families?', n.d; https://www.ethnologue.com/subgroups/indo-european [accessed April 23, 2022]

Hugh G. Evelyn-White, *Hesiod, Homeric Hymns, and Homerica*, William Heinemann, London, 1914

Brian Fagan, *Cro-Magnon: How the Ice Age gave birth to the first modern humans*, Bloomsbury Press, New York, 2010

Dimitra Fimi, *Tolkien, Race, and Cultural History: From Fairies to Hobbits*, Palgrave Macmillan, Basingstoke, 2008

Stephen Forbes, 'The oldest historical tree in the world', *Medium*, Dec. 26, 2016; https://medium.com/@StephenJForbes/the-oldest-historical-tree-in-the-world-566fcee32605 [accessed May 11, 2022]

Monica Gagliano, *Thus Spoke the Plant: A remarkable journey of groundbreaking scientific discoveries and personal encounters with plants*, North Atlantic Books, Berkeley, CA, 2018

Kashmira Gander, 'Road project in Iceland delayed to protect "hidden" elves', *Independent*, 23 Dec. 2013; https://www.independent.co.uk/news/world/europe/road-project-in-iceland-delayed-to-protect-hidden-elves-9021768.html [accessed Feb. 4, 2022]

Daniel Garber, 'Descartes, Mechanics and the Mechanical Philosophy' in *Renaissance and Early Modern Philosophy*, John Wiley and Sons, Oxford, 2002

Wendy Garling, *Stars at Dawn: The forgotten stories of the women in the Buddha's life*, Shambhala Publications, Boulder, CO, 2016

Philippe Gignoux, 'Hell in Zoroastrianism', *Encyclopedia Iranica*, 2003; https://www.iranicaonline.org/articles/hell-i [accessed March 17, 2022]

David Graeber and David Wengrow, *The Dawn of Everything: A new history of humanity*, Farrar, Strauss and Giroux, New York, 2021

Group Think, 'Lorde's *Ao Mārama*: Behind the songs', *The Spinoff*, Sept. 9, 2021; https://spinoff.co.nz/atea/09-09-2021/lordes-te-ao-marama-behind-the-songs [accessed June 5, 2022

Guardian staff and agencies, 'Happy the elephant is not a person, says court in key US animal rights case', *Guardian*, 15 June, 2022; https://www.theguardian.com/us-neews/2022/jun/14/elephant-person-human-animal-rights-happy [accessed July 17, 2022]

Herman Gunkel and H. Zimmer, *Creation and Chaos*, Vandenhoeck and Ruprecht, Göttingen, 1895

David L. Haberman, *People Trees: Worship of trees in northern India*, Oxford University Press, Oxford, 2013

John Hall, *Ernest Gellner: An intellectual biography*, Verso, London, 2011

Matthew Hall, *Plants as Persons: A philosophical botany*, SUNY Press, Albany, NY, 2011

——, *The Imagination of Plants: A book of botanical mythology*, SUNY Press, Albany, NY, 2019

Hyoung S. Ham, 'Manipulating the memory of meat-eating: reading the Lankāvatāra's strategy', *Journal of Indian Philosophy* 47, (2019): 133–53

Yuval Noah Harari, *Sapiens: A brief history of humankind*, Penguin Random House, London, 2015

——, *21 Lessons for the 21st Century*, Penguin Random House, London, 2019

——, 'The Actual Cost of Preventing Climate Breakdown', TED Talk, 2022

Peter Harrison, *'Religion' and the Religions in the English Enlightenment*, Cambridge University Press, Cambridge, 1990

Graham Harvey, *Animism: Respecting the living world*, Columbia University Press, New York, 2nd ed., 2017

Leonie Hayden, '"I'm beginning a journey": The inside story of Lorde's surprise mini-album in *te reo Māori*', *The Spinoff*, Sept. 9, 2021; https://thespinoff.co.nz/atea/09-09-2021\lorde-interview-maori-lyrics-solar-power [accessed June 5, 2022]

Jeremy W. Hayward, 'Scientific Method and Validation' in *Gentle Bridges: Conversations with the Dalai Lama on the sciences of the mind*, Shambhala, Boulder, CO, 2001

Alberto Hernando, D Villuendas, C. Vesperinas, *et al.*, 'Unravelling the size distribution of social groups with information theory on complex networks', *The European Physical Journal B* 76, (2009): 87–97

André Ferdinand Herold, *The Life of the Buddha*, trans. P.C. Blum, A. and C. Boni, New York, 1927

Israel Hershkovitz, Gerhard W. Weber, and Mina Weinstein-Evron, 'The earliest modern humans outside Africa', *Science* 359 (6,374), (2018): 456–9

Jacob Hill, 'Environmental consequences of fishing practices', n.d; https://www.environmentalscience.org/environmental-consequences-fishing-practices

Thomas Hobbes and Edwin Curley, *Leviathan: With selected variants from the Latin edition of 1668*, Hackett Publishing Company, Indianapolis, IN, 1994

John Holmes, 'Losing 25,000 to hunger every day', *UN Chronicle* (United Nations) 45 (3), (2009): 14–20

Mark Honigsbaum, *The Pandemic Century: A history of global contagion from the Spanish flu to Covid-19*, Penguin Random House, London, 2020 edition

Mark T. Hooker, *The Tolkienothēca: Studies in Tolkiennymy*, Lynfrawr, 2019

Masayuki Horie, Tomoyuki Honda, Yoshiyuki Suzuki, *et al.*, 'Endogenous non-retroviral RNA virus elements in mammalian genomes', *Nature* 463 (7,277), (2010)

Jack Hunter (ed.), *Greening the Paranormal: Exploring the ecology of extraordinary experience*, August Night Press, 2019

Christopher M. Hutton, *Linguistics and the Third Reich: Mother-tongue fascism, race, and the sciences of language*, Routledge, London, 1999

Hermann Jacobi, *Gaina Sutras Part 1: The Âkârânga Sûtra, the Kalpa Sûtra*, Clarendon Press, Oxford, 1884

Luc Janssens, Liane Giemsch, Ralf Schmitz, *et al.*, 'A new look at an old dog: Bonn-Obserkassel reconsidered', *Journal of Archeological Science* 92, (2018): 126–38

Ben Joffe, 'White Robes, Matted Hair: Tibetan Tantric householders, moral sexuality, and the ambiguities of esoteric Buddhist expertise in exile', PhD dissertation, University of Colorado at Boulder, CO, 2019

Janet Jones, 'Becoming a centaur', *Aeon*, Jan. 14, 2022; https://aeon.co/essays/horse-human-cooperation-is-a-neurobiological-miracle [accessed Aug. 25, 2022]

Jason Josephson-Storm, *The Myth of Disenchantment: Magic, modernity, and the birth of the human sciences*, University of Chicago Press, Chicago, IL, 2017

Walter S. Judd and Graham A. Judd, *Flora of Middle-Earth*, Oxford University Press, Oxford, 2017

Matthew Kapstein, *The Assimilation of Buddhism in Tibet: Conversion, contestation, and memory*, Oxford University Press, Oxford, 2000

Karma chags med, Gnas mdo, Gcod lugs char 'bod snyan rgyud yi get bkod par 'don bsgom phyag len dang bcas pa mthong bas don gsal, MW0LULDC313476, BDRC, n.d.

Donald R. Kelley, *Versions of History: From antiquity to the Enlightenment*, Yale University Press, New Haven, CT, 1991

Henry Ansgar Kelly, *Satan: A biography*, Cambridge University Press, Cambridge, 2006

Robin Wall Kimmerer, *Braiding Sweetgrass*, Penguin Random House, 2020

Robert Kirk, *The Secret Commonwealth of Elves, Fauns and Fairies*, Anodos Books, Dumfries and Galloway, 2018

Naomi Klein, *This Changes Everything: Capitalism vs. the climate*, Simon & Schuster, New York, 2014

Jakub Kocurek, 'Contemporary popular concepts of *Klu* and *Gnyan* as a result of ecological change', *Journal of Ethnology and Folkloristics* 7 (1), (2013): 19–30

Robin Korman, Lama Chonam, and Sangye Khandro, *The Epic of Gesar of Ling: Gesar's magical birth, early years, and coronation as king*, Shambhala Publications, Boulder, CO, 2015

Spencer Kornhaber, 'The pop star who's redefining the word "basic"', *Atlantic*, Aug. 19, 2021; https://www.theatlantic.com/culture/archive/2021/08/lorde-solar-power-review/619808/?utm_source=feed [accessed June 5, 20222]

Adam Krug, 'Buddhist medical demonology in the sutra of the seven buddhas', *Religions* 10 (4), (2019)

Maja Krzewińska, Gülsah Merve Kilinç, Anna Juras, *et al.*, 'Ancient genomes suggest the eastern Pontic-Caspian steppe as the source of western Iron Age nomads', *Science Advances* 4 (10), (2018)

Franciscus Bernardus Jacobus Kuiper, 'The basic concept of Vedic religion', *History of Religion* 15, (1975): 108–12

Alexander Kulik, 'How the devil got his hooves and horns: the origin of the motif and the implied demonology of 3 Baruch', *Numen* 60, (2013): 200

Elizabeth Kutter, Daniel De Vos, Guram Gvasalia, *et al.*, 'Phage therapy in clinical practice: treatment of human infections', *Current Pharmaceutical Biotechnology* 11 (1), (2010): 69–86

Bruno Latour, *Facing Gaia: Eight lectures on the new climatic regime*, Polity Press, Cambridge, 2017

Ma gcig Lab sgron, *Machik's Complete Explanation: Clarifying the meaning of* Chöd, trans. Sarah Harding, Snow Lion Publications, Ithaca, NY, 2013

Jennifer Larson, *Greek Nymphs: Myth, cult, lore*, Oxford University Press, Oxford, 2001

Will Lavin, 'Lorde tells fans her album is "going to take some time" after death of her dog Pearl', *NME*, 2 Nov. 2019; https://www.nme.com/news/music/lorde-tells-fans-album-is-going-to-take-time-death-dog-2563977 [accessed June 5, 2022]

Ursula K. Le Guin, 'The Critics, the Monsters, and the Fantasists', *The Wordsworth Circle* 38 (1/2), (2007)

Brian P. Levack, *The Oxford Handbook of Witchcraft in Early Modern Europe and Colonial America*, Oxford University Press, Oxford, 2013

Brian G. Levman, 'Cultural remnants of the indigenous peoples in the Buddhist scriptures', *Buddhist Studies Review* 30 (2), (2014): 145–80

——, *Pāli and Buddhism: Languages and lineage*, Cambridge Scholars Publishing, Newcastle upon Tyne, 2021

Charlton T. Lewis, 'Religio', *An Elementary Latin Dictionary*, Tufts University, Medford, MA, 1890; https://www.perseus.tufts.edu/hopper/text?doc=Perseus%3A1999.04.0060%3Aentry%3Dreligio

C.S. Lewis, *Lewis: Collected Letters, Volume 1: Family Letters. 1905-1931*, ed. W. Hooper, HarperCollins, London, 2000

Bruce Lincoln, 'The Indo-Europeans' myth of creation', *History of Religions* 15 (2), (1975): 121–45

Jigme Lingpa and Kangyur Rinpoche, *Treasury of Precious Qualities: Book One*, Shambhala Publications, Boulder, CO, 2010

Li Liu, Jiajing Wang, Danny Rosenberg, *et al.*, 'Fermented beverage and food storage in 13,000 y-old stone mortars at Raqefet Cave, Israel: Investigating Natufian ritual feasting', *Journal of Archeological Science: Reports* 21, (2018): 783–93

Sharon A. Lloyd and Susanne Sreedhar, 'Hobbes's moral and political philosophy', *Stanford Encyclopedia of Philosophy*, April 30, 2018

Lorde, 'Solar Power', *Solar Power*, comps. Ella Marija Lani Yelich-O'Connor and Jack Antonoff, Universal Music New Zealand, 2021

Hendrik Lorenz, 'Ancient Theories of the Soul' in *Stanford Encyclopedia of Philosophy*, 2009; https://plato.stanford.edu/entries/ancient-soul/ [accessed Oct. 3, 2021]

Robert J. Losey, Vladimir I. Bazaliiskii, Sandra Garvie-Lok, *et al.*, 'Canids as persons: early Neolithic dog and wolf burials, Cis-Baikal, Siberia', *Journal of Anthropological Archeology* 30, (2011): 174–89

Robert MacFarlane, *Underland: A deep time journey*, Penguin Random House, London, 2020

Christopher S. Mackay, *The Hammer of Witches: A complete translation of the* Malleus Maleficarum, Cambridge University Press, Cambridge, 2009

J.P. Mallory and D.Q. Adams, *The Oxford Introduction to Proto-Indo-European*, Oxford University Press, Oxford, 2006

Stefano Mancuso, *The Revolutionary Genius of Plants: A new understanding of plant intelligence and behavior*, Simon & Schuster, New York, 2017

——, *The Nation of Plants: A radical manifesto for humans*, Profile Books, London, 2021

Michael Marder, 'The life of plants and the limits of empathy', *Dialogue* 51, (2012): 259–73

——, 'Should animals have rights?', *Philosopher's Magazine* 62, (2013): 56–7; doi:10.5840/tpm20136293

Lynn Margulis and René Fester, *Symbiosis as a Source of Evolutionary Innovation*, MIT Press, Cambridge, MA, 1991

Joshua J. Mark, 'Mythology', *World History Encyclopedia*, Oct. 31, 2018; https://www.worldhistory.org/mythology/ [accessed Dec. 10, 2021]

Sarah Marshall-Pescini, Franka S. Schaebs, Alina Gaugg, *et al.*, 'The role of oxytocin in the dog–owner relationship', *Animals (Basel)* 9 (10), (2018)

Benjamin W. McCraw and Roberts Arp, *Philosophical Approaches to Demonology*, Routledge, New York, 2017

Ralph McInerny and John O'Callaghan, 'Saint Thomas Aquinas', *Stanford Encyclopedia of Philosophy*, May 23, 2014; https://plato.stanford.edu/entries/aquinas/ [accessed March 4, 2022]

Robin McKie, 'Loss of EU funding clips wings of vital crow study in Cambridge', *Guardian*, 28 May 2022; https://amp.theguardian.com/environment/2022/may/28/loss-eu-funding-crow-study-cambridge-brexit-corvid?fbclid=IwAR14BshF9Ldewb344EXTJ3Z

Mbqj-kStnQlkoCi6CMNy5YZAyLWKtncjdCU0 [accessed May 20, 2022]

G.yu thog yon tan Mgon po, *gDon nad gso ba bcos sgrig ma*, ed. Bya mdo klu byams rgyal, Mi rigs dpe skrun khang, Beijing, 2019

Steven J. Mithen, *After the Ice: A global human history, 20,000–5,000 BC*, Harvard University Press, Cambridge, MA, 2003

Marina Montesano, 'Horns, hooves and hell: the devil in medieval times', *National Geographic*, Nov. 2, 2018; https://www. nationalgeographic.co.uk/history-and-civilisation/2018/10/horns-hooves-and-hell-the-devil-in-medieval-times [accessed July 10, 2022]

Friedrich Max Müller, *The Upanishads, Part 2*, Clarendon Press, Oxford, 1879

Linda Nash, 'Beyond Virgin Soils: Disease as Environmental History' in *The Oxford Handbook of Environmental History*, Oxford University Press, Oxford, 2014

René de Nebesky-Wojkowitz, *Oracles and Demons of Tibet: The cult and iconography of the Tibetan protective deities*, Akademische Druck- und Verlagsanstalt, Graz, Austria, 1975

Mark Nielsen, Michelle C. Langley, C. Shipton, *et al.*, '*Homo neanderthalensis* and the evolutionary origins of ritual in *Homo sapiens*', *Philosophical Transactions of the Royal Society B*, June 29, 2020; http://doi.org/10.1098/rstb.2019.0424

Catherine Nixey, *The Darkening Age: The Christian destruction of the classical world*, Macmillan, London, 2017

Dan Nixon, 'The body as mediator: the phenomenology of Maurice Merleau-Ponty entwines us, via our own beating, pulsing, living bodies, in the lives of others', *Aeon*, Dec. 7, 2020; https://aeon.co/essays/the-phenomenology-of-merleau-ponty-and-embodiment-in-the-world [accessed Aug. 12, 2022]

Nonhuman Rights Project. 2022, quoted in *Guardian* staff and agencies, 'Happy the elephant is not a person, says court in key US animal rights case', *Guardian*, 14 June 2022; https://www.theguardian.com/us-news/2022/jun/14/elephant-person-human-animal-rights-happy [accessed July 17, 2022]

NRM, 'Microbiology by numbers', *Nature Reviews Microbiology* 9 (628), (2011)

Caitlin O'Connell, *Wild Rituals:10 lessons animals can teach us about connection*, community, and ourselves, Chronicle PRISM, San Francisco, 2021

Seán Pádraig O'Donoghue, *The Forest Reminds Us Who We Are*, North Atlantic Books, Berkeley, CA, 2021

Asko Parpola, *The Roots of Hinduism: The early Aryans and the Indus civilization*, Oxford University Press, Oxford, 2015

Hervey C. Peoples, Pavel Duda, and Frank W. Marlowe, 'Hunter-Gatherers and the Origins of Religion', *Human Nature* 27 (2016): 261–82

John W. Pilley, 'Border collie comprehends sentences containing a prepositional object, verb, and direct object', *Learning and Motivation* 44 (4) (2013): 229–40

Pliny the Elder, *Natural History: A selection*, trans. J.F. Healy, Penguin, London, 2004

Val Plumwood, *Feminism and the Mastery of Nature*, Routledge, London, 1993

——, 'Nature in the Active Voice', *Australian Humanities Review* 46, (2009)

Ross Pomeroy, 'Scientists have learned from cases of animal cruelty', *Real Clear Science*, 23 Jan. 2012; https://www.realclearscience.com/blog/2012/01/scientists-can-be-cruel.html [accessed May 10, 2022]

Joseph Poore and T. Nemecek, 'Reducing food's environmental impacts through producers and consumers', *Science* 360 (6,392), (2018): 987–92

Popham, Sajah Popham, *Evolutionary Herbalism: Science, medicine, and spirituality from the heart of nature*, North Atlantic Books, Berkeley, CA, 2019

Terry Pratchett, *Small Gods*, Gollancz, London, 1992

Stefan Riedel, 'Edward Jenner and the history of smallpox and vaccination', *Baylor University Medical Center Proceedings* 18 (1), (2005): 21–5

Hannah Ritchie, 'How much of the world's land would we need in order to feed the global population with the average diet of a given country?', *Our World in Data*, Oct. 3, 2017; https://ourworldindata.org/agricultural-land-by-global-diets [accessed July 20, 2022]

——, 'Half of the world's habitable land is used for agriculture', *Our World in Data*, Nov. 11, 2019; https://ourworldindata.org/global-land-for-agriculture [accessed July 20, 2022]

William Woodville Rockhill, *Life of the Buddha and the Early History of His Order: Derived from Tibetan works in the Bkah-hgyur and Bstan-hgyur*, Trübner & Co., London, 1884

Jean-Jacques Rousseau, *Discourse on the Origin of Inequality*, 1755; Hackett Publishing Company, Indianapolis, IN, 1992

Christopher B. Ruff, Erik Trinkaus, and Trenton W. Holliday, 'Body mass and encephalization in Pleistocene *Homo*', *Nature* 387, (1997):173–6

Edward P. Rybicki, 'The classification of organisms at the edge of life or problems with virus systematics', *South African Journal of Science* 86 (1990)

Carl Sagan, Lynn Margulis, and Dorian Sagan, 'Life' in *Encyclopaedia Britannica*, 5 Sept. 2022; https://www.britannica.com/science/life [accessed June 26, 2022]

Marshall Sahlins,'What kinship is (part one)', *Journal of the Royal Anthropological Institute* 17 (1), (2011), 2–19

Ian Sample, 'Neanderthals built mysterious cave structures 175,000 years ago', *Guardian*, 25 May 2016; https://www.theguardian.com/science/2016/may/25/neanderthals-built-mysterious-cave-structures-175,000-years-ago [accessed Oct. 10, 2021]

Geoffrey Samuel, *Civilized Shamans: Buddhism in Tibetan societies*, Mandala Book Point, Kathmandu, 1993

——, 'Spirit Causation and Illness in Tibetan Medicine' in *Soundings in Tibetan Medicine: Anthropological and Historical Perspectives, Proceedings of the Tenth Seminar of the International Association for Tibetan Studies*, *PIATS*, Oxford, Brill, Leiden and Boston, MA, 2003, pp.213–24

Sapienship, '2% more: the extra step to save our future'; https://www.sapienship.co/decision-makers/2-percent-more [accessed July 20, 2022]

Martin Saxer (director), *Journeys with Tibetan Medicine*, Docufactory, 2005

——, 'The Journeys of Tibetan Medicine' in *Bodies in Balance: The art of Tibetan Medicine*, Rubin Museum of Art and University of Washington Press, NewYork/Seattle, 2014, pp.246–56

Jill Schneiderman, *Anthropocene Feminism*, ed. Richard Grusin, University of Minnesota Press, Minneapolis, MN, 2017

Christina Scull and Wayne G. Hammond, *The J.R.R. Tolkien Companion and Guide*, HarperCollins, London, 2017

——, *The J.R.R. Tolkien Companion and Guide: Reader's Guide*, 2 vols, HarperCollins, London, 2017

Robert A. Segal, *Myth: A very short introduction*, Oxford University Press, Oxford, 2004

William Shakespeare, *The Tempest* in in *The Annotated Shakespeare*, eds B. Raffel and H. Bloom, Yale University Press, New Haven and London, 2006

Merlin Sheldrake, *Entangled Life: How fungi make our worlds, change our minds and shape our futures*, Penguin, London, 2021

——', Before Roots' in *This Book is a Plant*, Profile Books, London, 2022, pp.9–17

Brett Sholtis, 'When COVID deaths are dismissed or stigmatized, grief is mixed with shame and anger', *NPR*, Aug. 30, 2021; https://www.npr.org/sections/health-shots/2021/08/30/1011785899/when-covid-deaths-are-dismissed-or-stigmatized-grief-is-mixed-with-shame-and-ang?t=1661861693246 [accessed Aug. 29, 2022]

Suzanne Simard, *Finding the Mother Tree*, Allen Lane, London, 2021

Peter Singer, 'Animals' in *The Oxford Companion to Philosophy*, ed. Ted Honderich, Oxford University Press, Oxford, 1995

——, *Practical Ethics*, Cambridge University Press, Cambridge and New York, 2011

Marcello Siniscalchi, Serenella d'Ingeo, and Angelo Quaranta, 'Orienting asymmetries and physiological reactivity in dogs' response to human emotional faces', *Learning & Behavior* 46, (2019): 574–85

D.B. Smith, 'Mr. Robert Kirk's Note-Book', *The Scottish Historical Review* 18 (72), (1921): 237–48

Stephen D. Snobelen, 'Isaac Newton, heretic: the strategies of a Nicodemite', *British Journal for the History of Science* 32 (4), (1999): 381–419

Rebecca Solnit, 'Big oil coined "carbon footprints" to blame us for their greed. Keep them on the hook', *Guardian*, 23 August 2021; https://www.theguardian.com/commentisfree/2021/aug/23/big-oil-coined-carbon-footprints-to-blame-us-for-their-greed-keep-them-on-the-hook.

Isabelle Stengers, *In Catastrophic Times: Resisting the coming barbarism*, Open Humanities Press, Ann Arbor, MI, 2015

Martin D. Stringer, 'Rethinking animism: thoughts from the infancy of our discipline', *Journal of the Royal Anthropological Institute* 5 (4), (1999): 541–56

Snorri Sturluson, *The Prose Edda*, ed. and trans. Rasmus B. Anderson, University of Wisconsin, Madison, WI, 1879

——, *The Prose Edda by Snorri Sturluson*, trans. A.G. Brodur, Oxford University Press, Oxford, 1916

Martin B. Sweatman and Dimitrios Tsikritsis, 'Decoding Göbekli Tepe with archeoastronomy: what does the fox say?' *Mediterranean Archeology and Archeometry* 17 (1), (2018): 233–50

Martin Sweatman, 'The Younger Dryas impact hypothesis: review of the impact evidence', The University of Edinburgh, 2021; https://www.research.ed.ac.uk/en/publications/the-younger-dryas-impact-hypothesis-review-of-the-impact-evidence

Paul Taçon and Colin Pardoe, 'Dogs make us human', *Nature Australia* 27 (4), (2002): 52–61

Nyanaponika Thera, *The Roots of Good and Evil: Buddhist texts translated from the Pali with comments and introduction*, The Buddhist Publication Society, Penang, Malaysia, 1999

J.R.R. Tolkien, *The Lord of the Rings*, Allen & Unwin, London, 1954–5; HarperCollins, London, 1991, 1994

——, In Their Own Words: British Authors, BBC interview, 1968

——, *The Silmarillion*, ed. Christopher Tolkien, George Allen & Unwin, London, 1977

——, 'Myths Transformed' in *Morgoth's Ring*, ed. Christopher Tolkien, HarperCollins, London, 1993, pp.369–436

——, *The Book of Lost Tales*, 2 vols, HarperCollins, London, 2000

——, 'On Fairy-Stories' in *The Monsters and the Critics and Other Essays*, HarperCollins, London, 2006

——, *The Letters of J.R.R. Tolkien*, HarperCollins, London, 2006

——, *The Legend of Sigurd and Gudrún*, ed. Christopher Tolkien, HarperCollins, London, 2010

——, *On Fairy-Stories*, Expanded edition with commentary and notes, eds Verlyn Flieger and Douglas A. Anderson, HarperCollins, London, 2014

——, 'The Notion Club Papers' in *Sauron Defeated*, ed. Christopher Tolkien, HarperCollins, London, 2017

Jorge Torres, 'Plato's anthropocentrism reconsidered', *Environmental Ethics* 43 (2), 92021): 119–41

Mary Evelyn Tucker and Duncan Ryuken Williams, *Buddhism and Ecology: The interconnection of dharma and deeds*, Harvard University Center for the Study of World Religions Publications, Cambridge, MA, 1997

Edward B. Tylor, *Primitive Culture: Researches into the development of mythology, philosophy, religion, art, and custom*, John Murray, London, 1871

Gary Ulstein, 'Hobbits, Ents, and Dæmons: Ecocritical Thought Embodied in the Fantastic', *Fafnir – Nordic Journal of Science Fiction and Fantasy Research* 2 (4), (2009): 7–17

Marine Veits, Itzhak Khait, and Uri Obolski, 'Flowers respond to pollinator sound within minutes by inceasing nectar sugar concentration', *Ecology Letters* 22 (9), (2019): 1,483–92

Paul Veyne, *Did the Greeks Believe in Their Myths? An essay on the constitutive imagination*, trans. Paula Wissing, Chicago University Press, Chicago, IL, 1988

Nicholas Wade, *The Faith Instinct: How religion evolved and why it endures*, Penguin, New York, 2009

Stephanie Wakefield, *Anthropocene Back Loop: Experimentation in unsafe operating space*, Open Humanities Press, Ann Arbor, MI, 2020

Jeffrey D. Wall, Kirk E. Lohmueller, and Vincent Plagnol, 'Detecting ancient admixture and estimating demographic parameters in multiple human populations', *Molecular Biology and Evolution* 26 (8), (2009): 1,823–7

James H. Wandersee and Elisabeth E. Schussler, 'Preventing plant blindness,' *American Biology Teacher* 61 (2), (1999): 82–6

Shigeru Watanabe, Junko Sakamoto, and Masumi Wakita, 'Pigeons' discrimination of paintings by Monet and Picasso', *Journey of the Experimental Analysis of Behavior* 63 (2), (1995): 165–74

Linton Weeks, 'Recognizing the rights of plants to evolve', *NPR*, Oct. 26, 2012; https://www.npr.org/2012/10/26/160940869/recognizing-the-rights-of-plants-to-evolve?t=1653996829735 [accessed July 25, 2022]

Melinda Wenner, 'Humans carry more bacterial cells than human ones', *Scientific American*, Nov. 30, 2017; https://www.scientificamerican.com/article/strange-but-true-humans-carry-more-bacterial-cells-than-human-ones/ [accessed Nov. 11, 2021]

David R. Wessner, 'The origins of viruses', *Nature Education* 3 (9), (2010)

Gordon White, *Ani.Mystic*, Scarlet Imprint, London, 2022

Lynn White, 'The historical roots of our ecological crisis', *Science* 155 (1967): 1,203–207

Liz Williams, *Miracles of Our Own Making: A history of paganism*, Reaktion Books, London, 2020

Raymond Williams, 'Ideas of Nature' in *Problems in Materialism and Culture*, Verso, London, 1980

Peter Wohlleben, *The Hidden Life of Trees: What they feel, how they communicate – discoveries from a secret world*, trans. Jane Billinghurst, Greystone Books, Berkeley, CA, 2016

Andrew Woodhall, 'Addressing Anthropocentrism in Nonhuman Ethics: Evolution, morality, and nonhuman Beings', doctoral dissertation, University of Birmingham, Dept. of Philosophy, 2016

Yang Ga, 'Sources for the Writing of the *Rgyud Bzhi*, Tibetan Medical Classic', doctoral dissertation, Harvard University, Cambridge, MA, 2010

Ronit Yoeli-Tlalim, *ReOrienting Histories of Medicine*, Bloomsbury, London, 2021

Carl Zimmer, 'Ancient viruses are buried in your DNA', *New York Times*, Oct. 4, 2017; https://www.nytimes.com/2017/10/04/science/ancient-viruses-dna-genome-html [accessed Sept. 1, 2022]

INDEX

ABOUT THE AUTHOR

Erik Jampa Andersson is an interdisciplinary scholar, practitioner, and teacher of Tibetan Medicine, Buddhism, and Environmental History. He is a graduate of the Shang Shung Institute School of Tibetan Medicine, and has spent twenty years studying Tibetan Buddhist philosophy, practice, and ritual arts under numerous esteemed teachers around the world. With an academic background in religious studies and performance, Erik is currently completing an MA in History at Goldsmiths, University of London, where his research focuses on the critical intersection of ecology, mythology, and health.

Erik is the founder of Shrīmālā, where he offers consultations, coaching, and training on an array of topics to clients and students around the world. In addition to his Tibetan medical education, Erik has trained in Western herbalism, Taoist medicine, classical astrology, plant alchemy, and numerous other disciplines, and is particularly interested in the ways that such practices can impact our approaches to ecology and 'nature.' He also serves as a peer reviewer and contributor for TibShelf, specializing in translations of Tibetan medical literature. He lives in London with his husband and two cats.

For more information about Erik and his work, please visit:

www.shrimala.com

CONNECT WITH

HAY HOUSE

ONLINE

🌐 hayhouse.co.uk   **f** @hayhouse

📷 @hayhouseuk   🐦 @hayhouseuk

▶️ @hayhouseuk   ♪ @hayhouseuk

*Find out all about our latest books & card decks • Be the first
to know about exclusive discounts • Interact with our authors
in live broadcasts • Celebrate the cycle of the seasons with us
• Watch free videos from your favourite authors •
Connect with like-minded souls*

*'The gateways to wisdom and knowledge
are always open.'*

Louise Hay